Introduction to Natural Selection

JOHNSON, Clifford. Introduction to natural selection. University Park, 1976. 213p ill bibl index 76-8175. 12.50 ISBN 0-8391-0936-9. C.I.P.

CHOICE   JUNE '77

Biology

An attempt to quantify the factors involved in natural selection. Johnson almost loses the forest for the trees; he claims that biological importance need not be lost in mathematical symbolism and then proceeds to do just that. Admittedly, a rigorous mathematical analysis of the biological process of selection has a specific place in any study or evaluation. But an "Introduction" is not the place. It is in fact a review of the current literature pertaining to a quantitative analysis of natural selection. As such, it will appeal to graduate students and others wishing to update their knowledge of population genetics. The author does attempt to diversify his approach with examples from a number of different species (notwithstanding placing the smooth-bodied newt in the wrong genus). The bibliography is comprehensive and up to date. There seem to be no errors of fact, although the prose is so turgid and elliptical that it is not easy to properly evaluate all statements. As a reference volume for those studying natural populations, this book has a place on library shelves. But one wishes that the author had followed the example of the frequently quoted Dobzhansky and expressed his erudition with greater clarity.

Introduction to Natural Selection

Clifford Johnson, Ph.D.

Department of Zoology
University of Florida

UNIVERSITY PARK PRESS
Baltimore · London · Tokyo

UNIVERSITY PARK PRESS
International Publishers in Science and Medicine
Chamber of Commerce Building
Baltimore, Maryland 21202

Copyright © 1976 by University Park Press

Typeset by The Composing Room of Michigan, Inc.

Manufactured in the United States of America by Universal Lithographers, Inc., and The Maple Press Co.

Library of Congress Cataloging in Publication Data

Johnson, Clifford, 1932-
    Introduction to natural selection.

    Includes bibliographical references and index.
    1. Natural selection. I. Title.
QH375.J63   575.01'62   76-8175
ISBN 0-8391-0936-9

Contents

Preface

Natural selection is a process basic to essentially all biology and its study takes many directions, from the observations of field naturalists to the abstract, theoretical predictions of mathematicians. The process is so fundamental and outwardly simple that few introductory texts assess the actual evidence and fewer still describe the methods and assumptions required of its study. On one front, ecologists frequently allude to natural selection as the driving force of one process or another and frequently include it in their concepts and textbook treatments. The degree of complexity accompanying the operation of selection is, however, rarely covered at any depth in introductory ecology, and the problems and methods of documenting and measuring selection are even less frequently addressed. In a similar fashion, introductory courses of general genetics have such a rich body of material to cover that selection is rarely touched upon in detail. In yet another field, evolutionists have provided an array of updated texts on speciation covering the evolution of isolating mechanisms. The role of natural selection is again generously recognized, but the rigorous requirements for its actual measurement or operation are usually by-passed. The discipline bearing most directly on selection processes has come to be known as ecological genetics. Recent years have witnessed the appearance of several outstanding books identifying the concepts and progress in this area; however, the texts are addressed to either a general audience and give the problems in rather simple fashion, or they are addressed to advanced readers assumed to have extensive backgrounds. Again, a rigorous introductory level of natural selection is missing. In still another field, population genetics, the topic is logically expected; however, the subject matter is largely mathematical and existing texts concentrate on theoretical problems, rarely confronting the practicalities of studying selection in a natural population. The discipline also utilizes, not infrequently, mathematics so sophisticated that many biologists remain unacquainted with the precepts.

Nonetheless, methods have developed over a rather long period for measuring and documenting selection in natural populations. This book developed as an effort to bring together these methods in a fashion suitable for introducing students to the methods and assumptions biologists have found useful and necessary, and in addition, to provide a window to the research literature on this intriguing and complex area. At the same time, it has been possible to introduce the basic tenets of selection theory. This coverage assumes the reader is acquainted with general genetics and the mathematics used are generally the operations obtained from basic courses.

The organization is quite unlike that of existing books, and the concepts with supporting examples chosen for discussion in no way exhaust the whole picture. The illustrations or cited studies have been intentionally taken over a wide span to give a glimpse of the historical perspective on selection and are not presented as the sole crest of current knowledge. Many concepts and accompanying examples could clearly involve much lengthier discussion; however, if the introduction presented sufficiently interests a few serious readers so that they pursue the subject more deeply, its purpose will have been achieved.

I appreciate deeply the guidance obtained from students, colleagues, and correspondents who have played no little role in shaping the views discussed here, while all shortcomings are, of course, my own. I am indebted to the editors of the *Journal of Animal Ecology* and *Heredity* for permission to use, in part, tabular data from J. A. Bishop's study of a cline in *Biston betularia* (Chapter 5), and to redraft a figure from Kettlewell and Berry's study of a cline in *Amathes glareosa* (Figure 5-2). In addition, I acknowledge gratefully the permission of Doubleday and Co., Inc., to quote Charles Darwin's speculation on assortative mating in cattle on the Falkland Islands from their copyright 1962 Anchor Book Edition of the 1860 revision of *The Voyage of the Beagle*. Finally, it is a pleasure to acknowledge the invaluable assistance provided by the editors towards the book's final preparation.

Gainesville, Fla.
April, 1976

Introduction to Natural Selection

Chapter 1

Introduction

The concept of organic evolution or change of living forms through time is frequently identified directly with Charles Darwin. The idea of a change or origin of living forms can be traced, however, back through several centuries. In many ways, natural selection, though basic to essentially all biological processes, seems to be a simple system. In restrospect, one may well wonder why its recognition or acceptance was difficult. We must attempt to evaluate earlier interpretations in the light of what was then known of life and tempered by the values that the current philosophies or religions demanded.

The earliest notions of an evolutionary change dates back to Greek philosophers. Thales (640–546 B.C.) appears to have led the way by noting that natural explanations were perhaps superior to mythological ideas. His notions of life were quite simple. All life was thought to have come from water and, in fact, all life was made from water. Xenophanes (576–480 B.C.) recognized fossils as the remains of earlier life and noted the changes in distributions of the ocean. The first suggestion of a selection process appears to have come from Empedocles (495–435 B.C.). His era recognized fire, air, water, and earth, and he envisioned these materials to be acted on by two forces, love and hate. These forces in a competitive fashion formed animal parts that in turn united to form animals. Some were the natural forms we know, but others were two-headed monsters or other aberrant types. The battle of love and hate slowly led to the extinction of the monster forms by reducing their ability or reproduce. Empedocles thus envisioned a survival of the fit. Aristotle (384–322 B.C.)

culminated the list of Greek scholars and his influence lasted for many years. In the context of our subject, he believed that natural processes had a purpose by which the imperfect was gradually modified toward a perfect expression. He believed this perfecting principle was under an "Intelligent Design." Actually, Aristotle thought less of a selection process than Empedocles, but his teachings were also later banned in Europe by Church Doctrine.

The Roman era concentrated less on the origin of nature; however, the idea of survival of the fittest was apparently acceptable in their scale of values. The roman poet Lucretius (99–55 B.C.) gives a view of selection in "On the Nature of Things." The establishment of the Christian era in Western culture brought essentially all thought under direct control of the Church. The Greek legacy continued, however, with the philosophers of nature, Gregory of Nyssa (331–396) and Thomas Aquinas (1225–1274). These men continued to express Aristotle's views even though this activity was not allowed. A few outspoken writers like Roger Bacon (1214–1295) appealed for direct observation and open minds in looking for nature's secrets but the gates to such knowledge were closing. The writings of Arabic authors were beginning to appear in Europe, mainly by way of Spain. The Arabs had been largely influenced by the Greeks and were advocating their concepts about nature. A Church Council convened in Paris in 1209 and proscribed the works of Arabic writers in its region of influence, essentially all of Europe. Scientific questioning vanished, and the evolutionary concept did not re-emerge until the sixteenth century. Even then, all biological thought was overcast with the stamp of special creation. A broader view was not tolerated, as the Italian writer Giordano Bruno (1548–1600) discovered. He was burned at the stake.

Finally, during the seventeenth century, Francis Bacon in England and René Descartes in France survived their suggestion that species could change by nature acting on their variations, and that the idea of natural laws surpassed the dogma of creation in reaching an understanding of life. The gates to knowledge were open, the pathway to evolutionary concept was forming. The French scientist Georges Louis Leclerc de Buffon expressed a new view that variation from some original type involves both changes of improvement and of degeneration. Shortly thereafter, Erasmus Darwin, an English physician, writer, and Charles Darwin's grandfather, wrote that environmental changes could produce associated changes in animals. Bacon, Descartes, Buffon, and Erasmus Darwin expressed in their writings views close to the idea of the inheritance of acquired characters. These writers probably made little impact on the contemporary biologists of the era, since little or no response, either acceptance or rejection, resulted. Charles Darwin appears to have not held his grandfather's views in high esteem.

The next writer on evolutionary process elicited a huge response that continues to the present day. Most students in introductory biology are still firmly cautioned to avoid his notion of the inheritance of acquired characters. The author was Jean Lamarck. He compounded his theory by attempting to explain both the origin and the inheritance of variation (as is discussed in Chapter Two). Both the method of change and the concept of change itself appear to have been refuted by the contemporary biologists, particularly Lamarck's own countryman, Georges Cuvier, surprisingly enough a student of comparative anatomy.

Charles Darwin (1809–1882) arose against this background of evolutionary thought as an English naturalist with a very wide range of interest. During his career he pursued and published serious studies on such diverse topics as coral reefs, earthworms, barnacles, fertilization in plants, and the descent of man. Much of his interest in variation was, no doubt, stimulated by observations during a four-year cruise around the world visiting natural habitats with rich and diverse fauna. These experiences are well documented in *The Voyage of the Beagle*. All naturalists of his age devoted much effort to recognizing and classifying species, and Darwin was no exception. He appears, however, to have given more thought to the source of species than did most of his contemporaries. Darwin was also a well-read individual judging from his knowledge of existing literature. We do not know to what extent earlier views on evolutionary theory molded his opinions. We do know that he was particularly impressed with Charles Lyell's 1833 book, *Principles of Geology*, which described a long, gradual history of the earth's surface. He was also much impressed with Thomas Malthus' 1798 book, *An Essay on the Principle of Population*. He was systematically collecting facts for a book on variation in 1837, only one year after returning from the *Beagle*'s voyage. In the quest for such facts he continually encountered the problem of inheritance, for Mendel's classic study on genetics was not published until 1866, and apparently remained unknown to Darwin throughout his life. Darwin was deeply impressed by the various breeds man had been able to develop with domesticated animals. Improving desired features in domesticated forms appeared directly related to understanding natural improvements in wild species. This rationale led Darwin to personally undertake a project of breeding fancy pigeon varieties. If he had worked with an organism having less complex characters and larger families, he might have recognized the principles of inheritance. He clearly approached his breeding studies with proper quantitative zeal as he wrote in this context ". . . I have no faith in anything short of actual measurement and the Rule of Three," (from *Life and Letters of Charles Darwin*).

These experiences led him to recognize the process of artificial selection in domesticated animals. If such results could be obtained in domesticated animals, why, with enough time, could not changes of a similar magnitude occur in wild species? The process would be natural selection.

Darwin's correspondence records reveal that his concept and supporting facts were well assembled by 1838. The following 20 years allowed him to gather additional facts, but the actual reasons for his waiting so long to publish the study remain obscure. Some historians feel that Darwin anticipated the harsh response and chose to postpone it. Perhaps he would have waited still longer had not he received a manuscript from his contemporary, Alfred Russell Wallace. Wallace was unaware of the extent to which Darwin had pursued the problem of species, and was asking Darwin to comment on his manuscript before its publication. Wallace's manuscript was essentially Darwin's own view, and had been inspired by the same sources. Wallace, on learning of Darwin's long-standing work, granted him the publishing priority. However, both Darwin and Wallace jointly presented their concept to the Linnean Society of London on July 1, 1858. Darwin's book, *On the Origin of Species*, was then published on November 24, 1859. The first edition consisted of 1,250 copies, and all were sold on the first day. The book has clearly produced far-reaching controversial and influential effects on our society.

Darwin's principal contribution was injection of a mechanism that was not inconsistent with other biological properties, and was not directed by a supernatural design, to explain change. Basically, Darwin's impact can be traced to this sentence: "This preservation of favorable individual differences and variations, and the destruction of those which are injurious, I have called Natural Selection, or the Survival of the Fittest." His concept of selection, as advanced in the first edition of *On the Origin of Species* appeared at a time when the intellectual level of natural science was apparently optimal for the unveiling of a challenging and provocative concept. Fortunately so, for strong arguments rose against Darwin's proposal.

The first line of opposition rejected in toto the concept of natural selection, since it is not a directed change. This type of opposition to organic evolution by natural selection actually continues, in a diminishing degree, into recent times; however, in the cultural climate of the mid 1800's, this argument carried heavy influence. These attacks came as a disappointment to Darwin, a man of religious convictions. The right of free expression was nonetheless developed in England, and created a favorable setting for expounding a belief that, to many, struck at the foundations of Western culture. The defense led by Darwin's contemporaries against this rejection is quite probably given less than its deserved credit. His leading supporters were Thomas H. Huxley and Francis Galton, although they differed with him on biological particulars. Should they have remained silent, the impact of natural selection might well have been postponed. The controversy generated surely set the stage for the quick appreciation of Mendel's studies on inheritance. The rapid growth of genetics in England, and shortly thereafter in the United States, surely received much of its stimulus from the reactions to Darwin, reactions allowed to mature largely by the early defense raised by Huxley, Galton, and others.

The second line of opposition to Darwin's thesis came largely from other biologists, who questioned not natural selection, but rather its operation. Darwin believed that the inheritance of characters involved some sort of blending mechanism, and that selection, sorting over available possibilities in each generation, served to maintain variability. We now know that inheritance is not a blending system and that it actually operates to maintain variability, while selection may reduce it. The principal question in the late 1880's concerned, however, the type of variation responsive to selection. The fact that Mendel's classic study, published in 1866, was not either recognized or known until 1900 is now a textbook entry; however, the difficulties produced by this void of understanding encountered by early students of selection are less well-known. The most complete history of this colorful period in biology has recently been compiled by Provine (1971).

While Galton (Darwin's cousin) and Huxley came to the defense of natural selection, they challenged Darwin's view that selection could operate on the minute differences of many traits. They contended that discontinuous variations, or "sports," constituted the targets for selection. In an effort to describe his views, Galton developed the statistical process known as regression and applied the method to heredity, culminating in 1889 with publication of *Natural Inheritance*. Other scholars attracted to the study of selection agreed with Darwin that the minor differences or continuous variation seen in any one generation could, in time, be selected into distinct differences. The adherents of this position became known as Darwinists, or biometricians, and the leading representatives were Karl Pearson, contributing the mathematical expertise, and W. F. R. Weldon, a well-known biologist. The leading proponent for discontinuous variation following Huxley was William Bateson, who later became England's first scholar of Mendelian inheritance. The exchanges of opinion in this era often occurred in emotion-charged meetings, or in published articles attempting to discredit the work of others. Fisher (1930a), attempting to retrace the rationale from a publication of Bateson, many years later wrote ". . . a work, which owed its influence to the acuteness less of its reasoning than of its sarcasm . . ." In an effort to resolve these controversies, an "Evolution Committee of the Royal Society" was established in 1897 and, in order to have all views represented, Pearson, Weldon, and Bateson were included as members. Little progress was made by the Committee, but controversial eras often stimulate progress. The journal *Biometrika* was born as a result of the conflicting opinions. Bateson, serving as a referee for a paper submitted by Pearson to the Royal Society of London, recommended against its publication. As a result, Pearson and Weldon initiated the new journal as an outlet for their studies.

The arrival of Mendelian inheritance, followed shortly by de Vries' concept of mutation, was initially seen as almost overwhelming support for the proponents of discontinuous variation. These concepts also opened a new avenue to the experimental study of variation. The knowledge of inheritance began a rapid advance. In 1909 and 1910, Nilsson-Ehle in Sweden and E. M. East in the United States demonstrated that continuous variation was subject to laws of inheritance similar to those governing discontinuous traits, and the gap separating biometricians and Mendelians began to fade. In 1908, Hardy in England and Weinberg in Germany introduced the principle of genetic equilibria in random mating populations (although recent writers suggest Castle in the United States recognized this property as early as 1903).

The era was now witnessing a search for suitable experimental organisms, the verification of Mendelian ratios for numerous species, and appreciation of genetic possibilities in agriculture, etc. The study of natural selection seems to have taken a temporary recess while the foundations of genetics were sinking deep roots. The men who would renew that study were then being attracted to genetics and its potential for explaining evolutionary mechanisms. One such person turned from mathematics and astronomy, and in 1916, at the age of 26, submitted a paper on Mendelian inheritance to the Royal Society of London. The rationale and methods were somewhat of a departure from the biological convention of the day and the paper was rejected by the referees. Two years later, with the author's financial help, the paper, now an early classic in population genetics, was published by the Royal Society of Edinburgh. The author was R. A. Fisher, who 12 years later published what is probably the most often cited reference in the serious study of selection, *The Genetical Theory of Natural Selection.*

As the study of selection resumed, most workers accepted the premise of Fisher and others that selection operated at very low levels of intensity. This philosophy no doubt discouraged early attempts to measure the magnitude of selection in nature. An outstanding exception is found in the work of J. B. S. Haldane. Haldane, in 1924, appears to be the first worker to obtain a numerical solution for a selection coefficient as envisioned today. The main concern of many publications during the following 40 years of his life was the measurement and cost of natural selection. The names of Fisher, Haldane, and Sewall Wright are frequently identified with the development of the conceptual foundation of population genetics. Fisher and Haldane concentrated heavily on selection, although not always agreeing on details, and Wright brought attention to random change, inbreeding, and population size. Basically, Wright believed some traits or alleles, under certain population properties, could be neutral to selection, and his work laid the basis for an active area of current endeavor, namely, the effort to identify the unit, in genetic terms, necessary to perceive selection.

The students of evolution in the early 1900's were now equipped with genetic principles but were still probing the origin of species. The need for a population or biological concept of species, rather than the older typological approach, was recognized. In addition, they realized the need to identify those mechanisms that allow species to remain distinct. Darwin was largely concerned with a species' modification through long periods of time, or phyletic speciation. Later workers have largely addressed their studies to divergent speciation. In this process, one phyletic lineage bifurcates into two independent lines. The process of separating a common species into two simultaneously existing species presents a need for mechanisms that effectively isolate the two daughter species. Thus, the importance of reproductive isolation between existing species was recognized, and many studies have concentrated on identifying the isolation mechanisms between closely related species. The quest for the origin of species became a search for the origin of isolation mechanisms. The students of these processes all invoke natural selection, but their objectives rarely require direct documentation or measurement of selection.

Much is yet to be learned about speciation; however, attention has now turned more and more to the actual selection processes and to obtaining estimates for the magnitude of selection, a study frequently termed *ecological genetics*. Perhaps the most interesting result to date of these studies is the observation that many traits (the proportion remains unknown) experience far higher levels of selection loss, generally measured through differential survival, than earlier theoretical work envisioned. The frequency distribution for selective values is, of course, unknown, but if the early assumptions of selective values prove to be well below the actual mean, basic priorities in evolutionary theory could be restructured. The quest for selection evidence and magnitude, largely from natural populations, forms the subject matter of the following chapters.

The biological process of selection has now a long history and is rich in concept. Selection is the one process that is basic to all evolutionary biology. Consequently, students of this area view the details of a selection process seriously, since some systems will "work" and others will not. Discussions of selection can become embroiled in semantics. As an example, an individual may quickly be labeled as a group-selectionist, rather than as an adherent to the more preferred process of selection acting on individual variations, simply by a casual choice of words, rather than by any basic confusion. The difficulty can be attributed to a shorthand use of the language, much like the practice in genetics of simply saying red hair is recessive to black hair, rather than stating that the gene for the red hair trait acts in a recessive fashion when combined in the same genotype with genes determining the capacity for black hair. Beginning students can construe the shorthand usage to mean that the character impresses its action on the gene, and, of course, the speaker means nothing of the kind.

These difficulties vanish when symbolism is used to describe processes, though the biological importance need not be lost in mathematical sophistication. For this reason, and other more obvious uses, much of the study of selection requires some level of mathematical symbolism.

Chapter 2

Genetic Systems and Fitness

The principal effort to explain the change of animals through time and their adjustment to environmental requirements, before Darwin, came from the French zoologist Jean Baptiste de Lamarck. His views were presented in 1809 in his classic book, *Philosophie Zoologique*, wherein he advanced the concept of the inheritance of acquired characters using rather extreme examples and brought down a storm of ridicule. The theory was actually rather consistent with much of the knowledge of his era, and his use of an incorrect theory of inheritance often overshadows the fact that he led the pioneering work in formulating the first comprehensive explanation of biological change through time.

Both Lamarck and Darwin recognized the importance of variation but differed on its origin and potential for inheritance. Lamarck surmised that variation resulted from relative degrees of use and disuse, and that use led to greater expression for a trait, so it was acquired. The characters with this improved, acquired expression possessed a greater tendency to be transmitted to the next generation. Darwin simply took the position that all variation had potential for being transmitted to the next generation; his theory did not require an explanation for the variation's origin. These features of Darwin's concept allowed his theory to fall later into full accord with genetics. Darwin's vision of a blending inheritance system led him to believe that variation must be preserved by selection, while the blending inheritance worked to reduce variation. We currently realize that

9

selection can reduce or preserve variation but for reasons quite different than Darwin's. Even so, Darwin's view of the selection process has withstood the test of time and much scrutiny. The development of his theory takes five steps:

1. Many more young individuals are produced each generation than live to form part of the next breeding generation. This feature was impressed on Darwin by Thomas Malthus, and can be seen to allow much mortality each generation, some of it being selective, without reducing the average population numbers.

2. A species possesses considerable variation between individuals. The origin of the variation was independent of its later destiny.

3. All, or most, of the variation had a tendency to be inherited, a fact which is unlike Lamarck's position that only traits reinforced or acquired by use were inherited.

4. Some of these variations bestow on their possessors a higher chance of surviving or reproducing than exists for individuals with alternate variations. This feature was specifically selection, and was recognized also by Darwin's contemporary, A. R. Wallace. Note also that the chances of surviving and reproducing vary between individuals.

5. Long periods of time could lead to far-reaching modifications in a species. Charles Lyell's concept of long, gradual changes in the earth's geological history played a large role in Darwin's confidence in this theory.

Natural selection, from Darwin's definition, consists of the "preservation of favorable individual differences and variations, and the destruction of those which are injurious." The two agents allowing this preferential preservation are differential survival or reproductive success, or both, between individuals. If ". . . differences and variations . . ." should have no genetic basis for their determination, no impact of selection is carried forward to the hereditary structure of succeeding generations; however, few variations are free of some genetic control, and many attributes are completely hereditary expressions over wide environmental differences. When selection exists, its impact is reflected in some degree by individuals in successive generations and is equivalent to evolution. Selection may thus be observed directly by individual differences in survival and reproduction, or by changes in genetic structure of populations from successive generations. The observed changes in the genetics of successive generations owing to selection cannot always be partitioned into survival and reproductive effects. Namely, differential survival and reproduction translate into genetic effects without identifying the selection mechanism. Much of the evidence of selection, as seen in following examples, exists without a knowledge of the selection agent. The evolutionary significance of selection consists, however, of changes produced in a population's hereditary material; consequently, its study is closely related to the properties of the genetic systems. The theoretical properties of genes in populations are well covered in several works (Li, 1955; Crow and Kumura, 1970). The following review treats the basic properties required to understand selection. The necessary extensions of these properties are later developed relative to specific types of evidence or estimates of selection.

NEUTRAL EQUILIBRIA

Genes in populations are scored by their frequencies, and alleles combine into homozygous and heterozygous combinations in a manner determined by the pattern of matings or the mating system. A single autosomal locus of two alleles, A and a, has genotypes AA, Aa, and aa that may be symbolized by D, H, and R, where $D + H + R = 1.0$. If these symbols represent genotype numbers in a random sample, N, then the frequency of A, usually denoted by p, is:

$$\frac{2D + H}{2N}$$

since each D individual carried two A alleles and each individual has two possibilities of its occurrence. The frequency of a, denoted by q, is then $1-p$. If D, H, and R are given as proportions, than $p = D + 1/2\,H$ and $q = 1/2\,H + R$. If the basic mating pattern is the random system where alleles combine relative to their frequencies, then Table 1 illustrates the consequences of one generation of such mating. For the two-allele locus, with genes A and a having frequencies of p and q, respectively, the Hardy-Weinberg principle observes that alleles combine randomly forming genotype frequencies, p^2 for AA, $2pq$ for Aa, and q^2 for aa where $p^2 + 2pq + q^2 = 1.0$. These frequencies develop in only one generation of random mating and remain through successive generations if no disturbing influence occurs. The maintenance of this stability can be observed as follows. Among the three genotypes, six mating combinations exist, although where the two genotypes differ, these combinations exist in two forms giving a total of nine arrangements. These combinations are given in Table 2. The frequency for each of the six mating combinations appears in the second column and again proportionate parts of this value are given to the resulting offspring for each mating combination following Mendelian segregation. The algebraic sum of mating frequencies is seen to equal one (recalling that $p + q = 1.0$) and the F_1 genotype frequencies are again p^2,

Table 1. Establishment of equilibrium due to random mating

Type of mating	Frequency of mating	Offspring		
		AA	Aa	aa
$AA \times AA$	D^2	D^2		
$AA \times Aa$	$2DH$	DH	DH	
$Aa \times Aa$	H^2	$1/4\,H^2$	$1/2\,H^2$	$1/4\,H^2$
$AA \times aa$	$2DR$		$2DR$	
$Aa \times aa$	$2HR$		HR	HR
$aa \times aa$	R^2			R^2
Total	1.0	$(D + 1/2\,H)^2$ p^2	$2(D + 1/2\,H)(1/2\,H + R)$ $2pq$	$(1/2\,H + R)^2$ q^2

Table 2. Mating and offspring frequencies in a random mating population relative to a single gene locus

No.	Mating type	Frequency	Offspring		
			AA	Aa	aa
1	$AA \times AA$	p^4	p^4		
2	$AA \times Aa$	$2 \times 2p^3q$	$2p^3q$	$2p^3q$	
3	$AA \times aa$	$2 \times p^2q^2$		$2p^2q^2$	
4	$Aa \times Aa$	$4p^2q^2$	p^2q^2	$2p^2q^2$	p^2q^2
5	$Aa \times aa$	$2 \times 2pq^3$		$2pq^3$	$2pq^3$
6	$aa \times aa$	q^4			q^4
	Total	1.0	$p^4 + 2p^3q$ $+ p^2q^2 = p^2$	$2p^3q + 4p^2q^2$ $+ 2pq^3 = 2pq$	$p^2q^2 + 2pq^3$ $+ q^4 = q^2$

$2pq$, and q^2 for AA, Aa, and aa, respectively. Thus, regardless of the D, H, and R values, the p and q values give in one generation the frequencies of p^2, $2pq$, and q^2, respectively, and, once established, these values remain through successive generations. Among the offspring, the $(2D + H)/2N$ expression may be written as $(2p^2 + 2pq)/2$ which simplifies to p, stressing that gene frequencies have not changed, only genotype frequencies, and these values change only if D/N, H/N, and R/N do not equal p^2, $2pq$, and q^2 respectively.

Note, however, that several D, H, and R sets may possess the same value of p. Compare for instance $D = 200$, $H = 300$, and $R = 400$ with $D = 100$, $H = 500$, and $R = 300$. The value of p for both sets is $0.38\overline{8}$, yet the p^2, $2pq$, and q^2 values for a sample of 900 are approximately 136, 428, and 336, respectively. The latter genotype distribution reflects a population having a $p = 0.38\overline{8}$ mating at random while the former two D, H, and R sets, while each having a $p = 0.38\overline{8}$, are not consistent with a random mating explanation. Consequently, one p and q set may exist for many D, H, and R sets. Note also that the maximum value of H/N or $2pq$ under random mating is 0.5, occurring when $p = q = 0.5$. The p^2, $2pq$, and q^2 genotype frequencies are the so-called equilibrium frequencies and reflect a neutral equilibrium. If the locus is represented by three alleles, A, a^1, a^2 with frequencies of p, q, and r respectively, the equilibrium genotype frequencies are again obtained from expanding the expression $(p + q + r)^2$.

If genes at more than one locus are considered simultaneously, the gamete frequencies become the unit used above as gene frequencies. Two

Table 3. Frequency of gamete contributions in an equilibrium population for two gene loci and mating at random

Genotype	Frequency	Gametes			
		AB	Ab	aB	ab
$AABB$	p^2r^2	p^2r^2			
$AABb$	$2p^2rs$	p^2rs	p^2rs		
$AaBB$	$2pqr^2$	pqr^2		pqr^2	
$AaBb$	$4pqrs$	$pqrs$	$pqrs$	$pqrs$	$pqrs$
$AAbb$	p^2s^2		p^2s^2		
$Aabb$	$2pqs^2$		pqs^2		pqs^2
$aaBB$	q^2r^2			q^2r^2	
$aaBb$	$2q^2rs$			q^2rs	q^2rs
$aabb$	q^2s^2				q^2s^2
Total	$p^2(r+s)^2$ $+ 2pq(r+$ $s)^2 + q^2 =$ $(r+s)^2 =$ $(p+q)^2 = 1$	$p^2r(r+s)$ $+ pqr(r+s)$ $= pr(p+q)$ $= pr$	$p^2s(r+s)$ $+ pqs\,(r+s)$ $= ps(p+q)$ $= ps$	$pqr(r+s)$ $+ q^2r(r+s)$ $= qr(p+q)$ $= qr$	$pqs(r+s)$ $+ q^2s(r+s)$ $= qs(p+q)$ $= qs$

loci with alleles of A,a and B,b existing with frequencies of p, q and r,s, respectively, will at equilibrium occur in frequencies of pr, ps, qr, and qs for gametes AB, Ab, aB, and ab respectively. The frequencies of each genotype, at equilibrium, are the products of appropriate frequencies for each locus. For instance, $AABB$ occurs at a frequency of p^2r^2. When the genotypes exist in equilibrium frequencies, the gamete frequencies necessary for maintaining the equilibrium genotype frequencies are reproduced each generation, just as is the gene frequency under the single allele system. This property is illustrated in Table 3. These equilibrium gamete frequencies are, however, not developed in one generation of random mating unless the unlikely case of $p = q = r = s$ exists. The approach to equilibrium is affected by the extent to which frequencies differ from equilibrium values when random mating begins and by possible reduction of recombination due to linkage. These properties are discussed below in the context of linkage disequilibrium. Considering more than two loci simultaneously by this method can become tedious, and the statistical methods of quantitative or biometrical genetics are generally applied. If the locus is sex-linked, the nature of sex-linked inheritance is imposed on the system. The value of p changes between the sexes as the equilibrium value is approached. Equilibrium is also not obtained in a single generation of random mating. While the equilibrium value of p is similar for both sexes and the total population, the genotype frequencies differ, equaling p^2 and p, for instance, in the homogametic and heterogametic sex, respectively. The property of equilibria for sex-linked genes is discussed below in relation to selection on such loci.

The equilibrium frequencies of genotypes are not related to the mode of gene action. For instance, in dominant gene action we observe phenotypic frequencies of $p^2 + 2pq$ and q^2 in the two-allele system, or $p^2 + 2pr + 2pq$, $q^2 + 2qr$, and r^2 as one possible form of dominant action in a three-allele system. Among other things, this effect of gene action complicates the calculation of gene frequencies. The information in a sample decreases as the number of observed or scorable genotypes decrease, which can affect the accuracy of estimates for p, q, etc. Gene frequency estimates are nonetheless required in many selection studies, and various expressions have been derived for that purpose. Cotterman (1954) and Cook (1971) summarize procedures for p and q estimates and the following examples illustrate selected cases used in later discussions. The estimate of p for two-allele loci without dominance was given above as $(2D + H)/2N$, and where dominance obscures the distinction of D and H, an estimate of q may be taken from $\sqrt{R/N}$. This usage requires the assumption that D, H, and R are in equilibrium proportions, while estimating q by $(2R + H)/2N$

has no such limitation. Dominance generally exists for one member of an allelic series and the value of recognizing separate genotypes is obvious. The most direct way of partitioning dominant phenotypes into homozygotes and heterozygotes is by breeding tests, but this procedure is often difficult for large numbers. Cotterman (1954) has derived a helpful expression whereby estimates are possible with only a portion of dominant types identified. For instance, assume a sample of 108 individuals is available composed of 88 $M-$ and 20 mm types. Thirty-six of the 88 $M-$ are identified by breeding tests to genotype and the following categories now exist:

Class	$M-$	MM	Mm	mm	n
Observed No.	$52 = d$	$14 = e$	$22 = f$	$20 = g$	108

Cotterman's expression for the frequency of m is:

$$\frac{-(2e + f) + \sqrt{(2e + f)^2 + 8n\,(f + 2g)}}{4n}.$$

For the above set of values, the frequency of m is estimated as 0.4324.

In the case of three alleles with frequencies of p, q, and r, six genotypes exist. Depending on the gene action, at least three, four, or six phenotypes may be expressed. Six phenotypes correspond directly to the six genotypes and no difficulty in estimating frequencies develops. Gene action giving four phenotypes distributes genotypes in the following way:

Class	I	II	III	IV	n
Genotypes	$p^2 + 2pr$	$q^2 + 2qr$	$2pq$	r^2	1.0
Observed Nos.	214	208	151	57	630
Expected Nos.	212.6	206.6	152.8	57.9	630

The sum of classes II and IV is $q^2 + 2qr + r^2$ or $(q + r)^2$. Therefore, $1 - \sqrt{(q + r)^2} = p$. The sum of classes I and IV is $p^2 + 2pr + r^2$ or $(p + r)^2$, thus $1 - \sqrt{(p + r)^2} = q$. Now, $1 - (p + q) = r$, or $r = \sqrt{IV}$. The actual estimate usually employs the so-called Bernstein correction, applied as follows:

Classes	Observed Nos.	Frequency (f)	$\sqrt{f}$	First Estimate
II & IV	265	0.42063	0.64856	$p' = 0.35144$
I & IV	271	0.43016	0.65586	$q' = 0.34414$
IV	57	0.09047	0.30078	$r' = 0.30078$

$$0.99639$$

$$1 - 0.99636 = 0.00364 = d;\ d/2 = 0.00182$$

Adjusted Estimates:

$$p = p'(1 + d/2) \qquad\qquad = 0.3521$$
$$q = q'(1 + d/2) \qquad\qquad = 0.3447$$
$$r = (r' + d/2)(1 + d/2) \qquad = \underline{0.3031}$$
$$\qquad\qquad\qquad\qquad\qquad\qquad 0.9999$$

The expected numbers for each phenotype may then be entered for comparison; for example, $(p^2 + 2pr)$ 630 = 212.6, etc. The observed and expected values may be questioned for a significant departure from equilibrium values. The commonly used method is the chi-square test. Applying this test to gene frequency data involves fewer degrees of freedom than generally used. Usually, the degrees equal the number of phenotypes minus the number of alleles, in the above case, one. Deviations in this example are clearly not significant. Where gene action produces fewer than four phenotypic classes for three segregating alleles, some technique, such as breeding tests, is desirable for recognizing sufficient classes for a comparison of observed and expected distributions, although estimate methods exist, as shown in Table 4. The procedures for estimating gene frequencies with the most commonly encountered cases are tabulated in Table 4. The expressions are taken largely from Cotterman (1954) and Cook (1971).

NONRANDOM MATING SYSTEMS

Mating systems deviate from the random pattern in two basic ways. One form involves assortative mating based on phenotypically similar individuals. Positive and negative assortative mating consists of preferential matings between similar and dissimilar phenotypes, respectively. The second form is inbreeding, where matings between genetic relatives occur more frequently than random combination predicts. If only one gene locus determines the phenotype, positive assortative mating leads to high frequencies of homozygosity in few generations, but the effect decreases rapidly as the number of gene loci increase (Mather, 1973). Negative assortative (disassortative) mating reduces the homozygote proportions below values expected with random mating, but the effect is small when the number of loci exceeds one. Evidence of such mating systems exists in some natural populations but their contribution to observed gene frequencies is unclear. The process of imprinting by young on the parents' phenotypes is a likely setting for such an effect and is discussed by Seiger (1967). A genetically endowed basis for mating preference was found in male ischnuran damselflies for their choosing between the dimorphic

Table 4. Autosomal gene frequencies[a]

I. Two alleles; all genotypes recognizable.

Classes	AA	Aa	aa	Sample
Observed Nos.	d	e	f	n

$$p = (2d + e)/2n; \quad q = (2f + e)/2n; \quad \text{Var} = pq/2n$$

II. Two alleles; dominance.

Classes	$A-$	aa	Sample
Observed Nos.	d	e	n

$$p = 1 - q; \quad q = \sqrt{e/n}; \quad \text{Var} = (1 - q^2)/4n$$

III. Two alleles; partial classification of dominants.

Classes	$A-$	AA	Aa	aa	Sample
Observed Nos.	d	e	f	g	n

$$q = \frac{-(2e + f) + \sqrt{(2e + f)^2 + 8n(f + 2g)}}{4n}$$

$$\text{Var} = \frac{q(1 - q^2)}{2n\,(k - kq + 2q)}; \quad k = \frac{e + f}{d + e + f}$$

IV. Three alleles, dominant series.

Classes	A^1-	A^2-	aa	Sample
Genotypes	$p^2 + 2pq + 2pr,$	$q^2 + 2qr,$	r^2	1.0
Observed Nos.	d	e	f	n

$$p = 1 - \sqrt{\frac{e + f}{n}}; \quad \text{Var} = \frac{p(2 - p)}{4n}$$

$$q = \sqrt{\frac{e + f}{n}} - \sqrt{\frac{f}{n}}; \quad \text{Var} = \frac{q(2 - q[1 - p])}{4n\,(1 - p)}$$

$$r = \sqrt{\frac{f}{n}}; \quad \text{Var} = \frac{1 - r^2}{4n}$$

V. Three alleles, incomplete dominant series. (See example in text and Cotterman (1954) for Var estimates.)

Sex-linked gene frequencies

VI. Two alleles; all genotypes recognizable.

	Homogametes				Heterogametes		
Classes	AA	Aa	aa	Sample	Ay	ay	Sample
Observed Nos.	d	e	f	n	g	h	n'

$$p = \frac{2d + e + g}{2n + n'}; \quad q = \frac{2f + e + h}{2n + n'}; \quad \text{Var} = 2pq/2n + n'$$

---

[a]Cotterman (1954) and Cook (1971) also discuss cases where estimates of four or more alleles are desired within one series.

females, a system in which prior experience plays no role (Johnson, 1975). The literature contains few documented cases of the effects of assortative mating over long periods of time; however, Darwin (1860) gives an early description of a case concerning cattle on the Faulkland Islands that very possibly resulted from assortative mating. French colonists in 1784 brought a variety of breeds of cattle to the islands. During later years, control of these cattle waned and they became established as wild herds. When Darwin visited the islands in 1834, several distinct herds had separated, each characterized by specific traits and each occupying specific regions. Darwin gives the observations as follows:

> In colour they differ much; and it is a remarkable circumstance, that in different parts of this one island, different colours predominate. Round Mount Usborne, at a height of from 1,000 to 1,500 feet above the sea, about half of some of the herds are mouse- or lead-coloured, a tint which is not common in other parts of the island. Near Port Pleasant dark brown prevails, whereas south of Choiseul Sound (which almost divides the island in two parts), white beasts with black heads and feet are the most common: in all parts black, and some spotted animals may be observed. Capt. Sullivan remarks that the difference in the prevailing colours was so obvious, that in looking for the herds near Port Pleasant, they appeared from a long distance like black spots, whilst south of Choiseul Sound they appeared like white spots on the hillsides. Capt. Sullivan thinks that the herds do not mingle; and it is a singular fact, that the mouse-coloured cattle, though living on the high land, calve about a month earlier in the season than the other coloured beasts on the lower land. It is interesting thus to find the once domesticated cattle breaking into three colours, of which some one colour would in all probability ultimately prevail over the others, if the herds were left undisturbed for the next several centuries.

Inbreeding also develops homozygosity, potentially to a high degree. The rate depends on the closeness of relatives serving as parents, such as selfing, full-sib, double-first cousin, etc., in decreasing order of effectiveness. Inbreeding effect may be measured by reduction of heterozygotes below values predicted by random mating and expressed by the inbreeding coefficient, F. The inbreeding coefficient expresses the probability that two alleles at a locus are identical. The identity means the alleles arise from a common ancestral allele in a common parent and thus they are also identical in function. Thus, two homozygotes for a functionally similar allele need not involve identical alleles. Alleles that are identical through inbreeding are identified as identical by descent, or ibd. The coefficient may express the probability of identical alleles at a locus developed in one generation of random mating for a given population size (discussed in a later context), or the probability of a specific individual by using common ancestors in its pedigree. Both approaches assume that prior inbreeding was absent, which is of course unlikely for many cases. A simple measure of F, in a population where H is the observed percentage of heterozygotes is:

$$F = \frac{2pq - H}{2pq}.$$

Thus, $F = 0$ at random mating, and increases to 1 with selfing. References to these nonrandom possibilities for mating appear below in various contexts but two effects may be noted here. Inbreeding and positive assortative mating lead to no change in gene frequency, only genotype frequencies. Negative assortative mating has an extreme example in the case of self-sterility alleles, where a minimum of three alleles exist for a single locus. Such alleles interact to produce changes in both gene and genotype frequencies. The changes move frequencies toward equilibrium values of $0.33\overline{3}$ for both genes and genotypes (Rasmuson, 1961). Disassortative mating is occasionally termed outbreeding and, unlike positive assortative mating or inbreeding, requires a preferential restriction of some genotypes from mating to maintain the disassortative effect. This requirement introduces a selection component, thereby explaining the change in gene frequencies not seen in other mating systems. Disassortative mating is occasionally reported where self-sterility factors are not involved; however, such mating patterns are generally too weak to suggest much influence on gene frequencies.

A second effect of nonrandom mating concerns the changes of genotype frequencies. The genotype is an attribute of individuals, and inbreeding, for instance, produces a larger number of homozygous targets for selection to act upon than observed with random mating. This way, inbreeding could accelerate the effect of selection acting against a homozygote. Inbreeding is also associated with a decline in viability and fertility for some groups, the so-called inbreeding depression. This effect has important consequences in applied genetics but is rarely observed or identified in natural populations.

Consequently, nonrandom mating, particularly inbreeding, can influence the frequencies of traits and thus its recognition is important. If an F_1 sample is collected immediately after any one mating period (in practice impossible), then, assuming mating was random and no differential survival of gametes occurred, the genotype frequencies should be in equilibrium values for their parents' p and q values at the time of their mating. Beyond the time of zygote formation, a sample may differ from these equilibrium values because of causes other than nonrandom mating, such as differential survival of zygotes. If mating was random, each generation starts with equilibrium values from the previous generation's p and q values for single loci, and these values may persist throughout the age of the generation. Actually, as mentioned below, differential survival can exist during one generation in moderate intensity without being recognized by our tests. For these reasons, biologists sometimes assume differential gamete and zygote survival to be unlikely and collect samples well after zygote formation, compute p and q values directly from the samples, and compute Hardy-Weinberg expected phenotype numbers for comparison with observed numbers, as in the case above involving three alleles and four phenotypes. Of course, the older the age of the individuals composing the sample, the higher the likelihood of differential zygote survival affecting it.

Mantel and Li (1974) develop a variation of this test by noting the following:

$$2pq = 2\sqrt{p^2 q^2}; H = 2\sqrt{DR}, \text{ if equilibrium exists, or } H^2/4 - DR = 0.0.$$

They derive several estimates of variance for the left-hand term but the age structure of the sample still requires the same consideration noted above. Significant departures from equilibrium values in a single generation should be associated with other independent evidence before assuming that mating is nonrandom. Parsons (1967, 1973) reviews some experimental methods used with mating behavior that could assist decisions.

A selection pattern operating against the heterozygote could also lead to the observation that $H^2/4 - DR \neq 0$. The same effect may also result where all mating is random and no selection evidence exists. For instance, if D, H, and R of a sample are 0.38, 0.44, and 0.18, respectively, the Mantel-Li expression, $H^2/4 - DR$ is −0.02, rather than zero. The explanation could be that several separate population samples are being pooled. This event is only likely if the investigator failed to recognize heterogeneity of the habitat as the basis for separate populations. If habitats differ, then the magnitude of selection most likely differs from one site to the other, leading to different gene frequencies in each area. On this basis, the population in each area may be considered distinct.

The study area giving the above data can be imagined as consisting of five adjacent subdivisions, A through E, each differing in genetic properties, as outlined below:

Subdivision	P	q	p^2	$2pq$	p^2
A	0.8	0.2	0.64	0.32	0.04
B	0.7	0.3	0.49	0.42	0.09

C	0.6	0.4	0.36	0.48	0.16
D	0.5	0.5	0.25	0.50	0.25
E	0.4	0.6	0.16	0.48	0.36
Sums	3.0	2.0	1.90	2.20	0.90
	p^2		$2pq$	q^2	

Observed	$\Sigma p^2/5 = 0.38$	$\Sigma 2pq/5 = 0.44$	$\Sigma q^2/5 = 0.18$
Expected	$\bar{p}^2 = 0.36$	$2\bar{p}\bar{q} = 0.48$	$\bar{q}^2 = 0.16$
Difference	$+\,0.02$	$-\,0.04$	$+\,0.02$

Observed $\quad q = \bar{q} = \bar{p}\bar{q} + \bar{q}^2 = 0.22 + 0.18 = 0.40.$

The observed q is actually a mean q, as is the q^2. When the area was sampled without regard to the subdivision, the expected D, H, and R taken from the computed gene frequency, the mean $\bar{q}$, have frequencies of 0.36, 0.48, and 0.16, respectively. Note the observed H, 0.44, is lower than the expected H, 0.48. Yet, each population, A, B, etc., mates randomly with no selection effect. Furthermore, the observed departure from Hardy-Weinberg values involves an equal distribution of the heterozygotes' loss among the two homozygous classes, a feature of inbreeding. The likelihood of significant differences in gene frequencies existing between adjacent habitats that also remain unrecognized is probably low. The excess of the homozygous classes above expectation depends on the variation of q values in the pooled subdivisions of the habitat. Actually, the homozygous classes exceed the expectation, based on the $\bar{q}$ value, by an amount equal to the variance of q. The variance of q is obtained as follows:

$$[\Sigma f(q - \bar{q}^2)]/N = [\Sigma f(q^2 - 2\bar{q}q + \bar{q}^2)]/N \text{ or } [\Sigma fq^2 - 2\bar{q}\Sigma fq + \Sigma f\bar{q}^2]/N$$

$$= \Sigma fq^2/N - 2\bar{q}\Sigma fq/N + \Sigma f\bar{q}^2/N.$$

Since $\Sigma fq/N = \bar{q}$ and $\Sigma f = N$, then $V_q = \Sigma fq^2/N - 2\bar{q}^2 + \bar{q}^2$, or $\Sigma fq^2/N - \bar{q}^2$. Substituting the above values of $\Sigma q^2/5$ and $\bar{q}^2$, we obtain $V_q = 0.18 - 0.16$, or 0.02, the observed excess of each homozygous class. The result, a mimicry of inbreeding, is known as the Wahlund Effect.

SELECTION COEFFICIENTS AND FITNESS

Considerable mortality may occur during the life of a generation; however, if it is not differential between genotypes, their proportions at an advanced age will still be in equilibrium values. Likewise, reproductive success in a habitat may vary widely from one season to the next, but again, if it affects all existing types in a proportionately equal way, no differential success or selection results. Only a small part of total mortality or reproductive output need be differential to change the initial genotype frequencies. The genotype or phenotype experiencing maximal survival (or reproductive success) may be identified as the optimal type, with success of other classes expressed relative to this optimum. The conventional format for expression takes the fitness factor of the optimal class as 1, as shown below:

Genotypes	AA	Aa		aa	
Before-selection frequencies	p^2	$+ 2pq$	$+$	q^2	$= 1.0$
Fitness	1	1		$1 - s$	
After-selection frequencies	p^2	$+ 2pq$	$+$	$q^2(1 - s) = 1 - sq^2$	

In this example, dominant gene action produces an optimal phenotype of two genotypes and the recessive genotype experiences a drop in representation, expressed by the value s. Thus, the fitness factor ranges from 1.0 to zero and gives the after-selection genotype frequency. Fitness is generally symbolized by w. When w is <1.0, the magnitude of its reduction, s, ranging from zero to 1.0 is identified as the selection coefficient. Thus, $1 - s = w$ or $1 - w = s$. This convention treats s as having a positive sign, although negative s values may be used denoting a fitness exceeding 1. The sum of before-selection genotype frequencies times their fitness values gives the mean fitness, $\overline{W}$, following selection. In our example, $\overline{W} = 1 - sq^2$, and sq^2 is the mean selection coefficient, considerably less than s, since no selection existed for two genotypes.

A value used frequently with these expressions is the change of q between successive generations, or the difference equation, $\Delta q = q_2 - q_1$, where subscripts denote generations. In the example given, the after-selection frequency of a is $[pq + q^2(1 - s)]/(1 - sq^2)$ which simplifies to $[q(1 - sq)]/(1 - sq^2)$. If we assume no differential survival of gametes, this value will also be the before-selection q in generation 2. Thus, $\Delta q = [q(1 - sq)]/(1 - sq^2) - q$, or $[-sq^2(1 - q)]/(1 - sq^2)$. This expression gives a measure of gene frequency change in terms of s and the q value before selection.

Fitness may operate on genotypes in a number of ways. The most likely patterns are given in Table 5 with associated Δq expressions and a similar term for gametic selection. The algebraic derivation for these expressions are given in works like Li (1955), Cook (1971), etc. The expressions apply to an autosomal locus; however, if a sex-linked locus is considered, changes follow the gametic term for the heterogametic sex and the appropriate zygotic term for the homogametic sex. Where polyploidy

Table 5. Expressions for change of gene frequencies under different patterns of selection

Fitness patterns

Zygotic selection	AA p^2	Aa $2pq$	aa q^2	Δq
Gene Action				
Complete Dominance, Selection against aa	1	1	$1-s$	$\dfrac{-sq^2(1-q)}{1-sq^2}$
Incomplete Dominance, Selection against Aa and aa	1	$1-s$	$1-2s$	$\dfrac{-sq(1-q)}{1-2sq}$
Complete Dominance, Selection against AA and Aa	$1-s$	$1-s$	1	$\dfrac{sq^2(1-q)}{1-s(1-q^2)}$
Heterozygous Disadvantage	1	$1-s$	1	$\dfrac{sqp(2q-1)}{1-2spq}$
Heterozygous Advantage, Overdominance or Heterosis	$1-s_1$	1	$1-s_2$	$\dfrac{pq(s_1p-s_2q)}{1-s_1p^2-s_2q^2}$

Gametic selection	A p 1	a q $1-s$		$\dfrac{-sq(1-q)}{1-sq}$

exists, a given s leads to smaller Δq values (Mather, 1973). The case of selection operating on a unit of two or more loci is discussed below in different contexts. In all fitness patterns of Table 5, continued selection progressively reduces the frequency of genes existing in genotypes having fitnesses <1.0, except in the case of heterozygous advantage. Here both alleles enjoy maximal fitness in the same genotype and selection may operate without changing p and q values. In other words, Δq will equal zero, with selection in operation after p and q values come to equilibrium values. These values are often useful when expressed in terms of s or w. From Table 5, note that Δq for heterozygous advantage equals:

$$\frac{pq(s_1p-s_2q)}{1-s_1p^2-s_2q^2}.$$

A Δq of zero due to q or p equaling zero is no different from other fitness patterns, but a zero value also results when $s_1p - s_2q$ equals zero. This expression may be manipulated as follows:

$$s_1p = s_2q$$

$$s_1 p + s_1 q = s_2 q + s_1 q$$

$$s_1 (p + q) = q(s_2 + s_1)$$

$$\frac{s_1}{s_2 + s_1} = q$$

Similarly, the equilibrium value of p may be given as $s_2/(s_2 + s_1)$. If the fitness values for the three genotypes are expressed as a, b, and c, the following terms are available:

Genotypes	AA	Aa	aa
Before-selection frequencies	p^2	$2pq$	pq^2
Fitnesses	a	b	c
After-selection frequencies	ap^2	$2bpq$	cq^2

At equilibrium the following proportion exists:

$$\frac{p}{q} = \frac{ap^2 + bpq}{bpq + cq^2}.$$

Algebraically, $q = (b - a)/(2b - a - c)$, with q in terms of fitness values. This expression has become known as Fisher's Expression relative to equilibrium values. When gene frequencies are stable because of such balancing selection forces, the condition is described as a non-neutral equilibrium or balanced polymorphism. The p and q values before and after selection are similar, but the D, H, and R sets for the two periods are different. If only successive before-selection values are available, the selection pattern, if any, may be difficult to verify, a situation discussed below relative to the non-neutral equilibria.

ESTIMATING s AND w

The difficulty in estimating total, net fitness over a full generation will become more apparent as the following methods and assumptions are discussed. Partial fitness values comprise the great majority of our data, and when they are used this property must be kept in mind. The total, net s value over a full generation may be quite small and result from the summation of several large plus and minus components. Large selection coefficients found relative to only one life cycle stage may reflect only part of the total fitness. If the net s or $\bar{s}$ values are replaced by partial-generation values, then the relative intervals of selection become important. If coefficients of two or more traits are to be compared, the relative period of their operation may be as important as the magnitude of s or the frequency of suboptimal phenotypes. The point is made below, in the discussion of cryptic coloration, that some traits contribute to fitness for only a small percent of an individual's life span. Traits contributing to fitness over a long period may give no greater input than an attribute operating only during a short life-stage. For instance, 1,000 individuals experiencing an s of 0.05 per unit time have 750 survivors after five such intervals. One thousand individuals encountering an s of 0.25 over one comparable time interval also have 750 survivors. When comparing partial-generation coefficients for two or more traits of a species, the relative lengths of the life cycle concerned should be recognized. High s values per unit time are perhaps associated with attributes existing for only short periods. The available observations show reasonable agreement with the prediction. The high s values known (excepting disease) apply to adult insects having short adult expectancies, such as *Biston betularia*, discussed below.

If random mating exists and a single sample is taken at some age well beyond the zygote stage, a departure of genotype frequencies from the Hardy-Weinberg values for existing gene frequencies clearly suggests selection. However, it is possible to conclude erroneously that selection was absent, or that heterozygous advantage exists. For example, if $p = q = 0.5$, the following tabulation gives the basis for the single sample:

Genotypes	AA	Aa	aa
Before-selection frequency	0.25 +	0.50 +	$0.25 = 1.0$
Fitness	w_1	w_2	w_3
After-selection frequency	$0.25w_1$ +	$0.50w_2$ +	$0.25w_3 = \overline{W}$
Adjusted frequency	$\dfrac{0.25w_1}{\overline{W}}$ +	$\dfrac{0.50w_2}{\overline{W}}$ +	$\dfrac{0.25w_3}{\overline{W}} = 1.0$

If $w_1 = w_2 = w_3 = \overline{W}$, no selection has occurred, and $(w_1)(w_3) = w_2{}^2$. However, let w_1, w_2, and w_3 equal 1.0, 0.9, and 0.81, respectively. The adjusted after-selection genotype frequencies are:

$$
\begin{array}{ccc}
AA & Aa & aa \\
0.2770 & 0.4986 & 0.2244
\end{array}
$$

Therefore, $p = 0.5263$ and $q = 0.4737$. The genotype frequencies are in equilibrium for these p and q values. Likewise, $(w_1)(w_3) = w_2{}^2$. Thus, using the Hardy-Weinberg criteria, selection would erroneously be judged absent. The likelihood that w_1, w_2, and w_3 will by chance differ and still satisfy the above equality is low. If any one fitness differs from $\overline{W}$, and

$(w_1)(w_3) \neq w_2{}^2$, then selection must exist. However, if $(2pqw_2)/\bar{W} > 2pq$, then $w_2 > \bar{W}$ and $(w_1)(w_3) < w_2{}^2$, with heterozygous advantage appearing likely. However, several fitness sets again satisfy this inequality. For instance, if w_1, w_2, and w_3 are 1.0, 0.8, and 0.5, the inequality exists, but heterozygous advantage is absent. Clearly, a meaningful study of selection requires both before- and after-selection samples. The operations on genotype frequencies and fitnesses for two or more samples also require caution in their practical application. Ideally, the term "before-selection" refers to genotype frequencies formed at fertilization. In practice, the before-selection values come from young individuals sampled as soon after birth as circumstances permit, etc., or from laboratory-reared progeny of field-mated parents. This usage of before-selection assumes no gametic selection, and survival is to be measured during the development of this age-cohort. For many studies, survival may be the only component of fitness that is judged feasible to estimate. In such cases, the term "after-selection" is ideally the genotype proportions in adults from the same age-cohort at mating. The estimate of the frequencies is presumably best when inseminated females are collected and scored directly or by way of their offspring. A random sample of "adults" may not be reliable if their age structure differs noticeably from the age composition of breeders. Assuming ideal before-selection data, after-selection samples of inseminated females of the same age-cohort presumably give the closest approximation to any differential survival. Shorter periods may actually be desirable for some questions. If the after-selection sample represents a young age-cohort from the following generation, then the fitness estimate includes both survival and reproductive components without a general insight into their relative importance. The before-selection sample could also consist of adult breeders and the after-selection representatives come from a similar age group in the following generation, thus giving again both components of fitness. The validity of all after-selection samples will be compromised if migration has significantly influenced their properties. Each case must of course be judged independently.

In actual practice, samples are scored when the selection process is incomplete and a true estimate of net fitness is not obtained. Prout (1969) has examined the consequences and discussed procedures to minimize the errors. He identifies a pair of consecutive generations as a parent-offspring transition. When selection is incomplete at scoring, pre- and post-scoring components of fitness must be estimated, and he finds three transitions necessary. Also, if scoring in one transition occurs after all selection has been completed for both generations, the estimate is superior to scoring before-selection samples for both generations of one transition. Prout's study identifies limitations of field estimates and gives improvements for laboratory data. For students of natural populations, the main point is the recognition that fitness estimates are largely partial values of a generation, but for many questions these data may be quite adequate.

Fitness terms such as 1, 1, 1–s, etc., for p^2, $2pq$, and q^2, respectively, relate to individuals grouped by genotype and express relative success in survival and reproduction for one generation of selection, or success in only survival, often in a short, specified time. The fitnesses depicted in this way also imply that selection operates independently on different gene pairs at the value specified per genotype. The extent to which this assumption applies is discussed later; however, the algebraic operations apply equally to any units segregating and recombining as single factors. For instance, the inversions of *Drosophila pseudoobscura* are analyzed by expressing a given inversion frequency as p, etc.

Therefore, the fitnesses are relative, with the maximal value equaling 1.0 and the selection coefficient(s) being positive. This convention is generally followed; however, the relative fitnesses can be discussed using negative s values, as utilized in different contexts below.

Fitness may also be expressed as absolute or proportionate increase in the following ways. A typically used expression is:

$$\frac{\text{observed no. after-selection}}{\text{observed no. before-selection}}.$$

Absolute fitness is often measured in this way using direct numerical data and is only valid where after-selection samples consist of individuals directly surviving from before-selection samples. Data obtained from random collections have frequencies influenced by sample size, thus obscuring survival chance (O'Donald, 1971). For example, consider numbers in the following table:

Character	Before-selection	After-selection (1)	After-selection (2)
A	100 (0.50)	90 (0.5625)	56 (0.5656)
B	100 (0.50)	70 (0.4375)	43 (0.4343)

The frequency of character A in both after-selection samples is approximately 0.56 as could occur in two random collections. The relative fitness of B to A is 0.77, judged for both after-selection samples using methods described below; however, absolute fitness for B, by the above method, is 0.70 and 0.43 for samples 1 and 2, respectively. If absolute fitness is expressed proportionately as:

$$\frac{\text{observed \% after-selection}}{\text{observed \% before-selection or expected \% after-selection}}$$

the absolute fitness of B is seen to be 0.87 with both samples. This latter manner of handling absolute fitness has been used in earlier work by Kettlewell on recapture data discussed below. Relative fitness is similar, whether computed with numbers or with percents, using before- and after-selection data for within or between generation data.

Relative fitness also provides the best insight to selection. Ideally, it is found by finding the selection coefficient(s), s, from the difference equation for either Δq or Δq^2. If $q^2{}_B$ and $q^2{}_A$ represent the before- and after-selection frequencies, respectively, then:

$$q^2{}_A - q^2{}_B = \Delta q^2.$$

If we take the simple but common form of gene action involving dominance and the fitness set where the homozygous recessive genotype is at a selected disadvantage of s (directional selection), and express its fitness as $1 - s$, then the notation becomes:

Genotypes	AA		Aa		aa	
Before-selection	p^2	$+$	$2pq$	$+$	q^2	$= 1.0$
Fitness	1		1		$1 - s$	
After-selection	p^2	$+$	$2pq$	$+$	$q^2(1 - s) = 1 - sq^2$	

$$\Delta q^2 = \frac{q^2(1 - s)}{1 - sq^2} - q^2.$$

Rearranging, the right hand side of the equation gives:

$$\Delta q^2 = \frac{q^2(1 - s) - q^2(1 - sq^2)}{1 - sq^2} \quad \text{and thus the expression,}$$

$$\Delta q^2 = \frac{-q^2 s(1 - q^2)}{1 - q^2 s}$$

where the q^2 is the before-selection value. If we take a population for which $p = q = 0.5$ and also assume an s of 0.5, the frequencies are:

Genotypes	p^2	$2pq$	q^2
Before-selection	0.25	0.50	0.25
Fitness	1	1	0.5
After-selection	0.25	0.50	0.125
Adjusted frequencies		0.8571	0.1428

$\Delta q^2 = 0.1428 - 0.2500 = -0.1072$, the observed Δq^2.

If the $q^2{}_B$ and given s are placed in the Δq^2 expression, the expected Δq^2 value of -0.10714 is obtained. If the before- and after-selection q^2 values are available, rearrangement of the expression allows a solution directly for s by:

$$s = \frac{\Delta q^2}{-q^2(1 - q^2 - \Delta q^2)} .$$

Note that the use of Δq^2 in the above fashion takes only the before-selection sample to be in equilibrium for its gene frequencies. The after-selection value of q^2 was for adults in the same age-cohort, and the computed s thus concerns the survival component within one generation. The use of Δq usually applies to samples of two successive generations. If the fitness set involves two s values greater than zero, such as $1 - s_1, 1, 1 - s_2$, then two fitnesses are required relative to the maximal value of 1.0. Two Δq methods have been used for estimating such values. One approach requires two Δq equations solved simultaneously for the two unknowns. A second procedure used by Wright and Dobzhansky (1946) involves the mean fitness, $\bar{W}$. If p is expressed as $1 - q$, and the two selection coefficients as s and t, the following notation exists:

Genotypes	AA	Aa	aa

Before-selection $\quad (1 - q)^2 \ + \ 2pq \ + \ q^2 = 1.0$
Fitness $\qquad\qquad\quad 1 - t \qquad\quad 1 \qquad 1 - s$

After-selection $\qquad (1 - q)^2(1 - t) + 2pq + q^2(1 - s) = \bar{W}$
$\qquad\qquad\qquad\quad \bar{W} = 1 - sq^2 - t(1 - q)^2$

$$\Delta q = \frac{q(1 - q) + q^2(1 - s)}{\bar{W}} - q = \frac{q(1 - sq - \bar{W})}{\bar{W}}.$$

By rearrangement,

$$s = \frac{1}{q}\left[1 - \left(1 + \frac{\Delta q}{q}\right)\bar{W}\right].$$

By using this expression for s, the equation for t becomes:

$$t = \frac{1}{1 - q}\left[1 - \left(1 - \frac{\Delta q}{1 - q}\right)\bar{W}\right].$$

Therefore, an estimate for $\bar{W}$ allows a solution for the two selection coefficients. Wright and Dobzhansky develop an expression for $\bar{W}$ at the

$\Delta \bar{q}$ and $\bar{q}$ values from the generations in their data. The derivation involves the regression coefficient of Δq on q, $b_{\Delta q \cdot q}$, and $\bar{W}$ is given in reciprocal form as:

$$\frac{1}{\bar{W}} = 2 \left(1 + \frac{\Delta q}{q}\right)\left(1 - \frac{\Delta q}{1-q}\right) - 1 - b_{\Delta q \cdot q}$$

where Δq and q are $\Delta \bar{q}$ and $\bar{q}$. The solution assumes similar s and t values at all values of q, an unlikely situation. Several possible refinements of the estimates are also given.

These methods are tedious and have been used mainly for populations of *Drosophila*. Nonetheless, most selection studies require fitness estimates equivalent to values obtained by Δq methods and associated with an expression of variance for obtaining confidence limits. The "cross-product ratio" has been widely adopted to meet this need. The ratio, expressible in several forms, has a simple basis and is derived from adjusting two ratios into proportional equivalents. Its use is exemplified by the following table:

Character	Before-selection		After-selection		
	Number	Percent	Number	Percent	Expected
A	a_0	a_{po}	a_1	a_{p1}	a_e
B	b_0	b_{po}	b_1	b_{p1}	b_e
	N_0	1.0	N_1	1.0	1.0

Take two samples of 1,000 each, where $a_0 = 750$, $b_0 = 250$, $a_1 = 857$, and $b_1 = 143$. The following ratios exist:

$$\frac{a_0}{a_1} = \frac{b_0 w}{b_1} = \frac{750}{857} = \frac{250w}{143}$$

$$w = a_0 b_1 / a_1 b_0, \text{ or } w = 0.500$$

The term w is an adjusting factor applied to the before-selection frequency of the character that is proportionately less abundant in the after-selection sample. It expresses the relative fitness of B to A, taking A's fitness as 1.0. If the percentages, a_{po}, b_{po}, etc., are used, the same result is obtained. Percentages in the above examples are the same as in the previous Δq^2 example, where the s value was given at 0.5. Since $1 - w = s$, the example illustrates that results are comparable to the Δq^2 results for whatever time transpired between samples. The ratio has, therefore, two before- and two after-selection terms expressed as actual numbers or proportions, and the adjusting factor. The expected values, a_e and b_e, may be either numbers or percents, and are used of course with after-selection observed values such as:

$$\frac{a_{p1}}{a_{ep}} = \frac{b_{p1}}{b_{ep}w} \; ; w = \frac{a_{ep}b_{p1}}{a_{p1}b_{ep}}.$$

The term w, therefore, is used as a factor of before-selection or expected frequencies, and if the convention of a maximal fitness equaling 1.0 is adopted, it is a factor of the frequency observed to decrease. Using the ratios given above for absolute fitness, the relation of the two fitnesses can be seen:

$$\frac{\text{absolute fitness of B}}{\text{absolute fitness of A}} \text{ as } \frac{b_1/b_0}{a_1/a_0}, \text{ or} \frac{b_{p1}/b_{p0}}{a_{p1}/a_{p0}} = \frac{b_1 a_0}{a_1 b_0}$$

equals the relative fitness of B to A.

In the Δq method cited above from Wright and Dobzhansky (1946), they found t and s to equal 0.289 and 0.680, respectively. The maximal fitness of 1.0 occurred for the heterozygote and, accordingly, a balanced polymorphism was predicted. In such a system, p and q equal $s/(s+t)$ and $t/(s+t)$, respectively, or 0.7017 and 0.2982. The following values could, therefore, be predicted:

	p^2	$2pq$	q^2	
Before-selection	0.4924	+ 0.4185	+ 0.0889	= 1.0
Fitness	0.711	1.0	0.32	
After-selection	0.3501	+ 0.4185	+ 0.0284	= 0.7970
Adjusted frequency	0.43924	+ 0.52507	+ 0.03569	= 1.0

Using the cross-product ratio on these proportions, we obtain the following fitnesses:

$$\frac{2p_0 q_0}{2p_1 q_1} = \frac{q_0^2 w}{q_1^2} \; ; w = 0.3199, \; s = 0.680$$

$$\frac{2p_0 q_0}{2p_1 q_1} = \frac{p_0^2 w}{p_1^2} = w = 0.7109, \; t = 0.289.$$

Clearly, these values are comparable to the Δq solution illustrating the expression of two fitnesses relative to a third. Note that the p and q values are the same in the before- and after-selection samples, though the genotype frequencies differ. The cross-product ratio may utilize genotype frequencies for between generation analysis, where, of course, the expected frequencies in the second generation come from the p and q values of the first. The ratio technique allows analysis of within-generation data for phenotypic variability, while Δq procedures require an understanding of genetic inheritance. When computing fitness from the ratio, a rather large error may accumulate in the fitness factor if expected values are computed from p and q estimates that have been rounded. For instance, Wilson and Bossert (1971) give an example of fitness computation, $p52$, where p and q are rounded to 0.56 and 0.44. The fitness they obtain for the aa genotype is 0.77. If the p and q values are retained as 0.5562 and 0.4438, respectively, the fitness is found to be 0.7438 or 0.744. A difference of this magnitude, approximately one percent, may be important if confidence limits are computed as discussed below.

A number of experimental designs have been developed to measure differential predation, predator choice, etc. in controlled and rather artificial circumstances. Their analysis usually employs some modification of the cross-product ratio and the various designs have been summarized by Manly, Miller, and Cook (1972). Two modifications apply more directly to natural populations, and involve: 1) comparing samples from separate subsets of a population, and 2) incorporating results from successive ratios obtained from capture-mark-release-recapture techniques. The first exception is exemplified by comparing a predated group to the living population or a male population to the total population. This case is discussed below in regard to confidence limits for w. Manly (1972) states that random variation will affect proportions of two morphs even though both have the same probability of survival. To reduce this effect he derives a sampling procedure whereby successive recapture results can be pooled to reduce the random component. Samples large enough to have minimal effects yet consisting of successive recaptures seem unlikely for most species in natural settings. This error is probably best avoided by working toward larger samples that comprise a single ratio suitable for applying the following estimates of confidence limits.

The following table gives some values typical of actual samples:

Character	Before-selection		After-selection	
	Number	Percent	Number	Percent
A	315	0.4968	93	0.5636
B	319	0.5031	72	0.4363
Totals	634	0.9999	165	0.9999

Presumably no significant fitness difference will exist between characters if the before- and after-selection distributions are not different. The above numbers can therefore be evaluated in this regard with a 2X2 contingency table. The χ^2 (chi-square), using Yates correction, is 2.08, and for one degree of freedom, no significant difference seems to exist. B is, however, seen to fall in frequency and its relative fitness estimate is found to be 0.7645. Wolff (1955) derived a means of attaching confidence limits to the factor by expressing w as its $\log_e w$ value, and obtaining the variance of the log expression. For this example, $\log_e w$ is -0.26845. Wolff's expression for the variance is: $V_w = 1/a_0 + 1/a_1 + 1/b_1 + 1/b_0$, using the above ratio notation. For this example, the variance thus equals 0.03095, and 1.96 times the standard deviation is 0.3448. The antilogs of -0.23845 ± 0.3448 provides a 95 percent confidence range of 0.545 to 1.079 for w. Thus the range barely agrees with the χ^2 test by overlapping the maximum of 1.0. Cook (1971) derives another maximum likelihood estimate using numbers and proportions directly from the data. His expression is:

$$V_w = \frac{w(wb_{po} + a_{po})}{N_1(b_{po}a_{po})} \ .$$

The V_w equals 0.0144, and 1.96 times the standard deviation is 0.2352. Therefore, the 95 percent confidence limits are 0.765 ± 0.2357, or 0.5298 to 1.0002. The two methods give fairly similar intervals; however, it would be tempting to overlook the small overlap of 1.0.

If the before- and after-selection samples in the preceding table represent living and predated "populations" or similar separate groups, then Edwards (1965) shows that the ratio should be:

$$\frac{wb_0}{b_0 + b_1} = \frac{a_0}{a_0 + a_1}$$

where w equals 0.9463 for this example. Edwards proceeds to derive an estimate for confidence intervals following Wolff's method. The $\log_e w$ here equals -0.05518 and his variance expression is:

$$V_w = \frac{1}{a_0} + \frac{1}{b_0} - \frac{1}{a_0 + a_1} - \frac{1}{b_0 + b_1} \ .$$

The value here is 0.00129, and the standard deviation times 1.96 is 0.071. The antilogs of -0.05318 ± 0.071 give a range of 0.881 to 1.016. Again, the data grouped as separate sets fail to obtain significance.

Where χ^2 tests do not verify a difference in distribution, the 95 percent confidence limits of w will probably also be unacceptable.

RELATIONS OF Δq AND q

The documentation of w with such confidence limits depends, in addition to reasonably large samples, on the magnitude of Δq or Δq^2. Our tests are, therefore, largely insenstive to small coefficients and may also fail to recognize even large coefficients if q is near zero, 1.0, or an equilibrium point of a balanced polymorphism. The Δq values are small at each of these q values, as seen in Figures 1 and 2. The derivations of the Δq expressions used in those figures are discussed in Li (1955). High s values operating against a single genotype, as in Figure 1, can quickly move q to a low or high value. The likelihood of studying a natural population for such selection during the time that Δq is large enough to document is low. Note that Δq is between generations. In a balanced polymorphism, the Δq between generations remains near zero with low or high s values; however, the Δq^2 for the within-generation period may be rather large for high s values, and the effect is repeated each generation. Even here, if the selection coefficients are quite different, such as $s = 0.9$ and $t = 0.1$, the Δq of zero is at a q of 0.9, Figure 2. With dominance, the two morphs before selection would be at phenotypic frequencies of 0.81 and 0.19, close to the hypothetical second population discussed below.

Such equilibrium points increase the difficulty of documenting a given value of s. Consider the two following sets of hypothetical data, where Δ morph A is −0.060 and −0.023 in populations 1 and 2, respectively:

	Before-selection		After-selection	
	Numbers	Percent	Numbers	Percent
Population 1				
Morph A	500	0.50	444	0.440
Morph B	500	0.50	556	0.560
Population 2				
Morph A	900	0.90	877	0.877
Morph B	100	0.10	123	0.123

For both populations, w for $A \approx 0.80$ and $s \approx 0.20$. The following figures are taken from Cook's estimate of confidence limits. The 95 percent confidence limits of w for population 1 are 0.686 to 0.894, and for population 2 are 0.608 to 0.972. Both estimates of w are significant, but the sample sizes are larger than most field studies can assemble. If the after-selection sample sizes are reduced from 1,000 to 500, but retain the same percentages for morphs A and B, the results take the following form: The w and s for both populations are 0.79 and 0.21, and the 95 percent confidence limits for w of populations 1 and 2, respectively, are 0.638 to 0.932, and 0.528 to 1.042. The second population w is no longer significantly below 1.0, due to the smaller, proportionate effect the selection produced on the dissimilar initial frequencies. Variations studied in natural populations will infrequently concern s values much greater than 0.5, and

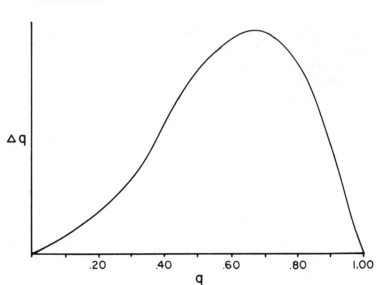

Figure 1. The curve for $\Delta q = -sq^2(1-q)/(1-sq^2)$ where s is constant. Note the low Δq values near q values of zero and 1.0.

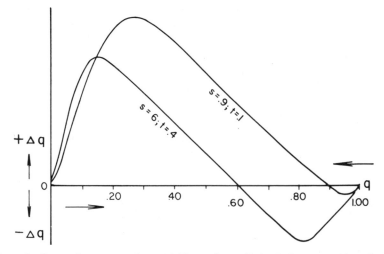

Figure 2. Curves for $\Delta q = pq(sp - tq)/(1 - sp^2 - tq^2)$ for indicated s and t values. Note: 1) the low Δq values near q values of zero, 1.0, and the intermediate equilibrium point, and 2) the shift away from $q = 0.5$ as the s and t values become progressively unequal.

confidence intervals are most likely to be used in identifying w's significantly less than one, rather than those exceeding zero.

The difference equations and cross-product method both may fail to recognize fitness differences if heterozygous advantage exists with lethality or sterility. For example, if $p = q = 0.5$, the before-selection genotype frequencies occur in a $1:2:1$ ratio. If both homozygous classes are lethal, the only breeders are heterozygotes where the $p = q = 0.5$ property remains. The before-selection genotype frequencies in the next generation are again $1:2:1$. If before- and after-selection samples in the same generation are available, such lethality may be detected, but if successive before-selection samples are used, the selection is missed. If the lethality is replaced by sterility, the selection is missed in both kinds of samples.

LIFE TABLES AND FITNESS

Differential reproductive success may exist without any differential survival. If s should exist only for differential reproductive success, the young and adult frequencies within a generation would not reflect any selection. Selection measured by Δq values between successive generation samples will usually include both components of fitness, and we may frequently have little knowledge of the relative importance of survival and reproductive output.

Reproduction is best recognized by life table data that follow an initial number, L_0, of individuals throughout life or indirectly approximate this treatment. The necessary life table entries are few, though often difficult to obtain. The age span of a generation must be known and divided into age groups, x, determined by particulars of the species in question. The number of individuals alive at the beginning of each age category is L_x, namely, those that have survived from the initial number observed at the first age, L_0. The midpoint of an age interval is termed the pivotal age. A survivorship curve is obtained by plotting the consecutive L_x values against the progressive units of age. Such curves are probably most accurate when computed from individuals observed throughout the full generation. Species with generation times approximating that of the observer require, of course, less direct methods. Life table procedures are given by several authors, such as Deevey (1947) and Krebs (1972). The reproductive data are obtained where L_x values concern survival in a female cohort only. In addition to age grouping and survival probabilities with age, the mean number of offspring (for some purposes, only female offspring) per female parent, relative to her age, is also required. This parameter is denoted m_x and the values are treated in the following fashion:

Age Group, x	Pivotal Age	L_x	m_x	$L_x m_x$
0–9	5.0	0.98	0.0	0.0
10–14	12.5	0.90	0.33	0.297
15–19	17.5	0.70	0.44	0.308
20–24	22.5	0.60	0.22	0.132
25–over		0.05	0.00	0.00

$$\sum_0^\infty L_x m_x = 0.737$$

For the age group 10–14 the probability of reaching the pivotal age of 12.5 is 0.90, and 3.3 females in 10 are expected to have a female offspring during this age period. The term, $\Sigma L_x m_x$ is denoted as R_0, the net reproductive rate, and in the example given, the female type measured is seen to fall short of replacing itself. If an R_0 value is obtained for each of several alternate phenotypes, the maximal value, R_0^{max}, can be taken as the optimal class with a fitness of 1.0. The relative values of other classes such as R_0^1, R_0^2, etc. are obtained by R_0^1/R_0^{max}, etc. These values measure the relative production of female offspring between recognized female classes. Actually, population analysts prefer to use the parameter identified as the innate capacity for increase, r_m. This measure of reproductive success also derives from life table data, and, in addition to R_0, an estimate of the mean generation time, G, is required. The value of G is $\Sigma L_x m_x X/R_0$ (Krebs, 1972). The estimate of r_m is then obtained from $\log_e(R_0)/G$. A number of population properties, such as the extent that generations overlap, can influence the accuracy of the estimate. The r_m parameter has received attention mainly from workers on theoretical considerations or ecologists who have not distinguished r_m values for separate phenotypes. Theoretically, a maximum r_m exists relative to the ideal environment and a phenotype's intrinsic abilities to multiply its numbers. Of course the ideal environment rarely, if ever, occurs and probably for only short periods. Thus, any computed r value will fall short of the theoretical maximum, but nonetheless, the actual r values achieved provide the necessary measure for comparing relative success.

Reproductive success in females measured by m_x reflects number of young, eggs, etc., produced per age and rarely is influenced by a failure to mate. In males, reproductive success can be measured by ability to mate, and Trivers' review (1972) concludes that the copulation number gives the best measure. For many species, some form of aggressive behavior accompanies male breeding. In this context, the most successful mating records correlate typically with greater body size, mobility, and past experience. Laboratory estimates for male success exist, as exemplified by Bateman's studies (1948) with *Drosophila melanogaster*. Efforts to collect similar field data working with the copulation number assumption are reviewed by Trivers (1972).

Occasionally, special components of fitness are recognized, namely, development time and mating ability. Actually, development time is a feature of the m_x curve with age and is that period required of a zygote to reach reproductive ability. If all other events remain equal, then the phenotype with the shorter development time obtains a head start and develops a higher net R_0 than an alternate morph with a longer development time. However, if the shorter development time is also accompanied by a lower life expectancy than accrues to the morph having a longer required development, then the maximal fitness depends on the full m_x curve. Thus, $\Sigma L_x m_x$ includes survival, reproductive output, and development time. The latter parameter alone may or may not play a major role in fitness. Mating ability, as indicated, is a property of more concern to males, measuring their fertility as the m_x entries serve for females, but assuming that sterility is insignificant. The component is also studied under the title of sexual vigor and, in this case, fertility differences of the male sex are often given as a case of sexual selection.

Net fitness is difficult to obtain from before- and after-selection samples, largely because such samples do not contain all of one generation's selection. Using methods designed largely for laboratory colonies, Prout (1971a and b) outlines a solution by estimating the components of fitness separately. The rationale of separate components has been used for some time with natural populations. If one component differs between types, an equal net fitness for the types is highly unlikely. In effect, selection has been verified by a partial fitness demonstration. In natural populations, the component most easily studied is survival, which may be measured during a fixed time interval or as the natural mean life expectancy. True fitness is the perpetuation of genetic material and a one-generation measure may be insufficient. If survival time and reproductive output for two types result in the same progeny number for each, they appear to enjoy equal fitness. However, if fertility of the offspring for one type is less than exists for the other, the result is equivalent to the proportion of the young that are sterile. Not only is the fitness of the semi-sterile young reduced, but also that of its parents. Inbreeding is the most likely process to introduce this effect.

MEAN LIFE EXPECTANCY AND FITNESS

While the estimate of r directly may be pursued along the lines suggested above, most studies seek a measure of increase in terms of e^{tr}.

Texts on population ecology develop expressions describing exponential increase of numbers in an environment free of competition. Such expressions are usually given as $N_t = N_0 \cdot e^{tr}$, where N_0 represents initial numbers, N_t gives numbers after time t, r is the innate rate of increase, and e is the base of natural logarithms. For our purposes, t is generation number and may be taken as one. If two morphs, A and B, of a species are under consideration, we have:

$$\frac{N_{At}}{N_{A0}} = e^{rA} \text{ and} \frac{N_{Bt}}{N_{B0}} = e^{rB}$$

where e^{rA} and e^{rB} are net rates of increase for A and B, respectively. The relative fitness of A to B for a full generation is then e^{rA}/e^{rB}. If appropriate numbers generated by these expressions were available, the result is a cross-product ratio:

$$\frac{(N_{At})(N_{B0})}{(N_{A0})(N_{Bt})} = e^{rA}/e^{rB}, \text{ or}$$

$$\frac{N_{A0}w}{N_{At}} = \frac{N_{B0}}{N_{Bt}}, \ w = \frac{(N_{At})(N_{B0})}{(N_{A0})(N_{Bt})}.$$

The value of t was taken as 1.0, but it could represent any chosen proportion of a generation. An exponentially increasing population is, however, an abstraction rarely existing, if ever. We need a comparable expression for populations with overlapping or discrete generations, and where numbers approximate an equilibrium. For life table data of such populations, the net rate of increase for a generation is:

$$\frac{1}{L_0} \int_0^\infty L_x m_x dx$$

following Birch (1948) and others. Here L_0 is again the initial number (usually given as females), L_x is the number alive from the original number at the beginning of age x, and m_x is number of offspring per female during x to $x + 1$. The probability of an individual being alive at age x is L_x/L_0. Relative fitness of two morphs could be measured by a ratio of expressions specific to morphs, as in the ideal condition above. In actual practice, life tables treat time, x, as discrete intervals; consequently, the net rate of increase is computed by summation rather than integration, and we have:

$$\frac{1}{L_0} \sum_{i=0}^{n=\infty} L_x m_x.$$

If no differences can be demonstrated for m_x of A and B, the parameter may be treated as a constant, and the expression is:

$$\frac{1}{L_0} \sum_{i=0}^{n=\infty} L_x.$$

Treating m_x as a constant between morphs is probably incorrect; however, without data otherwise, the fitness simplifies to the component of survival. Note that ΣL_x actually equals the total number of age units, e.g. days, survived. Thus, $\Sigma L_x/L_0$ equals the average number of age units survived, and the mean life expectancy at birth or from initiation of measurement, such as the emergence of adult insects from the larval-pupal period. Thus, $(1/L_0)\Sigma L_x$ equals e_0, the mean life expectancy. An estimate of this parameter for each morph would therefore give the relative survival component of fitness over the full-generation period. The e_0 value has the following relation to death rate, d. The death or mortality rate is the number that die in time x divided by the number alive at the beginning of time x, or:

$$d_x = \frac{L_x - L_{x+1}}{L_x}.$$

From this term we obtain another expression for L_x, or:

$$L_x = \frac{L_x - L_{x+1}}{d_x}.$$

If death rate is constant at all intervals of x, this expression is $L_x = (L_x - L_{x+1})\, 1/d$. Substituting this L_x expression into the $(1/L_0)\Sigma L_x$ expression for e_0 gives:

$$\frac{1}{L_0} \sum_{i=0}^{n=\infty} (L_x - L_{x+1})\, 1/d.$$

The summation of $L_x - L_{x+1}$ equals L_0, the number initially observed at birth. The expression reduces to $(1/L_0)(L_0/d) = 1/d$. By considering m_x constant for two morphs, the e_0 for each morph may be expressed as $1/d$, where d is the specific though constant death rate of each morph. Life tables are often constructed for females only; however, the assumption of reproductive equality for male morphs allows the expression to apply equally to males. Of course, life expectancies may well differ for the sexes. Where death rates are computed for only a portion of the life span, such as adults, and are assumed to be constant over the adult span, the adult life expectancy, e_a, is $1/d$ where d is now the adult death rate. This interval appears to be an ideal portion of a generation for measuring partial fitness. This rationale was followed by Fisher and Ford (1947) in developing their analysis of adult insect populations. Adults for many species have short life spans, and assuming a constant death rate over this period allows calculation of expected numbers closely matching observed data. Death rates may be obtained several ways. Fisher and Ford computed daily death rates by identifying, in an iterative fashion, the daily survival rate that produced expected numbers of survival days having the least total discrepancy from the observed number of survival days. Having computed this value, e_a then equals $1/d$, and where independent estimates of e_a exist, the values are reasonably close. The relative fitness of A to B becomes:

$$\frac{1/d_A}{1/d_B} = \frac{d_B}{d_A}.$$

If appropriate numbers of these d values exist, the expression is also equivalent to a cross-product ratio. If A and B are both represented by 100 individuals at time x and by 80 and 50 individuals, respectively, at time $x + 1$, then $d_A = (100 - 80)/100$ or 0.2, and $d_B = (100 - 50)/100$ or 0.5. The relative fitness of B to A is then 0.2/0.5, or 0.4. By cross-product ratio, the numbers dead at the end of an interval are compared, rather than the numbers alive. Without selection, 35 of each morph are expected to die (total deaths/2 since each started at equal numbers), thus:

$$\frac{\text{No. } A \text{ expected to die } (w)}{\text{No. } B \text{ expected to die}} = \frac{\text{No. } A \text{ observed to die}}{\text{No. } B \text{ observed to die}},$$

and $w = 0.4$. The w must be used as a factor of A rather than B since A, having the higher survival, had the lowest proportion lost. The summation expression for life expectancy, $(1/L_0)\Sigma L_x$, may overestimate that value if d, though constant, is high, say >0.2. A correction for this effect is given in a later chapter.

REPRODUCTIVE VALUES

Another measure of reproduction was derived by Fisher (1930a) and is known as the reproductive value. A separate value exists for each age and verbally is: the relative number of female offspring that remains to be born to each female of a specific age. The value is relative to expectation at birth. A summation expression for reproductive value, v_x, may be given as:

$$\frac{e^{rx}}{l_x} \sum_{y=x}^{\infty} e^{-ry} l_y m_y,$$

where e is the base of natural logarithms and r is the innate capacity for increase. This value, v_x, is computed for each female age concerned and the subscript y simply means all ages she can pass through from the point of measurement, x, to infinity. Derivation of the above expression is given by Wilson and Bossert (1971). Where generation time is short and age intervals proportionately large, an integral expression gives more accurate estimates. Reproductive value seems to have no advantage to R_0 or e^r in estimating fitness; however, the value does suggest some interesting features of selection. The curve of reproductive value with age, where known, is convex, having a maximum at an age well beyond birth. Thus, an individual's v_x increases with age from birth to some point in mid-life, after which v_x declines to zero. This curve shape is due to the high mortality rate typical of young age classes. If suboptimal classes are not affected by selection until the age of maximal v_x, then selection appears to generate a higher s value than if the selection acted at an earlier or later age; however, Hamilton (1966) questions this conclusion. An argument has occasionally appeared for the evolution of senescence by natural selection related to v_x. The rationale assumes that genes with high fitness during maximal v_x, but causing physiological deterioration at later ages, become fixed in the homozygous condition. It is not clear why genes with high fitness at one age need have negative fitness effects at a later age. Perhaps all genes develop a malfunction beyond a point in functional history. Hamilton (1966) interprets senescence as a selection response related more to fertility rates. Once maximal v_x begins to fall, for whatever reason, selection has a difficult time maintaining fitness with increasing age.

HETEROZYGOSITY AND SELECTION

The relative proportion of gene loci in an individual that is heterozygous, and thus segregating at meiosis, has been an energetically pursued parameter among evolutionists in recent years. Basic interpretations of evolution are at stake. If many loci are heterozygous and the resulting genotypes sensitive to selection, then the potentially large segregating genotype number will consist of many suboptimal types. The loss by selection is possibly very high. If selection coefficients per locus are modestly low, even 100 loci can generate a high total loss if each locus contributes independently to fitness. The problem is removed, at first sight, in three ways. The number of heterozygous loci may be low enough to involve only a few segregating alleles, or the selection coefficients may be exceedingly small, or selection may totally vanish and the resulting genotypes are neutral. These solutions are, however, not very compelling in the light of most experimental data. Recognizing individual loci, especially for traits with continuous variation, is difficult, and obtaining random samples from recognizable loci constitutes another problem. The advent of gel electrophoresis into genetic analysis allows the direct scoring of many alleles coding for enzymatic proteins. Lewontin (1974) describes the general rationale of this method. The randomness of alleles scored by the procedure relative to characters with a genetic determination is not clear, but the method appears to give the least biased estimate. Basically, to date, the average number of scored loci in an individual that are heterozygous is 10 percent, and an average of about 30 percent of a species loci is heterozygous (Lewontin, 1974). A very conservative estimate of loci number, even for a *Drosophila* species, is 10,000. If 10 percent of this number is heterozygous, the number of genotypes is unbelievably large, even if each locus has only two alleles. For instance, if 10 loci segregate for only two alleles each, 59,049 genotypes, 3^{10}, are possible in an F_2 generation. If such loci contribute independently to fitness, then the selective loss per locus must be extremely low if the population is going to successfully maintain itself. But, we find documented evidence of distinct fitness differences for factors segregating as single units. The proportion of such factors in individuals is of course unknown and doubtlessly varies, but many selection coefficients exceed 0.1, a value far too high to be an average for many independently operating loci. The third solution places most of the heterozygosity in a neutral category. Again, the evidence based on minimal assumptions points to a sensitivity to selection for most studies (Lewontin, 1974).

The most satisfactory reconcilation of data recognizes that the unit of selection is not, for most genes, single loci. Blocks of genes appear to be the unit where fitnesses of genotypes reflect some function of all concerned loci. Furthermore, the units appear to be largely integrated by way of linkage, whereby the number of segregating genotypes is drastically reduced without reducing allelic variation. This topic will be turned to again, but if we assume the presence of much polymorphism or heterozygosity and detectable selection, as most data suggest to be the case, we are then committed to a position for which fitness for most loci cannot be realistically considered independent of a number of other loci.

OTHER PATHWAYS TO Δq

The premise followed in this book is that the majority of genetic variation for the great majority of populations can at any time be predominantly explained by selection. This position is not accepted by all biologists.

Changes by way of mutation during short study periods are very limited, due to the low frequency of the rates. An average estimate of mutation for most loci is about 10^{-6} per generation. Microbes with short generation intervals and rapid growth may experience gene frequency change of a moderate magnitude through mutation, but such cases are largely beyond diploid animal populations, which are the concern of discussions here.

Changes by way of migration are real possibilities that must be judged by specific particulars of each case. A basic problem exists concerning this point. A population may be defined: as the inhabitants of a very definite circumscribed region; as a group receiving no more than some specified percentage input by migration; or as a group living in a region, large or small, having frequencies of a trait within a certain range, etc. Some definitions are convenient for purposes of calculations but have questionable biological validity, while other interpretations seem reasonable from biological viewpoints but pose serious difficulties for numerical treatment. The term "interbreeding" is often used in defining a population, but does this mean within a generation or by way of relatives geographically dispersed? Designing a measure of migration requires certain arbitrary definitions that confine realities to a minimum. Thoday (1974) recently stressed this point nicely in relation to disruptive selection. The difference in gene frequency along a distance gradient is usually very low over the mean dispersal distance of individuals of most species. Studies in a circumscribed area, regardless of the worker's concept of population, should show little effect from migration. Where frequencies differ by rather large values over short geographical distance, more concern for migration is required. The algebraic treatment of migration possibilities is covered by Cook (1971) and others. The expressions for migration and mutation are very similar; however, gene mutation is a single locus event while a single migrant event affects genes at every locus.

The likelihood of gene frequency changes associated with controversial opinion is *genetic drift*. The phenomenon operates as a sampling effect where random survival or combination of individuals for mating, or both, may deviate from the Hardy-Weinberg values and also override the effect of selection. The mathematics of drift are covered by Li (1955), Crow and Kimura (1970), and others. The principal factors concern population size, the value of q, the number of generations having a specific population size, and the magnitude of s occurring simultaneously. Sewell Wright was the first worker to suggest an evolutionary importance for drift, and the process is often called the Wright Effect. When drift operates, p and q shift randomly until the frequency of one allele reaches a value of one, and the allele concerned becomes fixed. The allele with the highest frequency when drift begins has the higher likelihood of fixation. Thus, given enough time, drift leads to homozygosity that may be nonadaptive, sometimes referred to as the decay of variability. Drift is calculated for its effect on p and q values under idealistic conditions, i.e., where population numbers are constant, inbreeding negligible, generations nonoverlapping, and each individual makes equal genetic contributions to each successive generation. Crow (1954) has considered such effects in detail for population size. If one or more of these restrictions are absent, then the drift likelihood relates to some number generally less than observed and the chance for drift increases. Correcting the observed population number for deviations from the ideal structure gives the so-called effective population size. When all corrections are known, an observed population of 300 breeding individuals may possibly lose heterozygosity at the same rate as an ideally structured, but much smaller, population, e.g. one of 50 individuals. The most easily assessed deviation is the sex contribution, reflected by the sex ratio. Correcting for this one deviation, as is generally done, gives an unknown overestimate of the effective population size. Other features of reproductive biology can actually increase the effective size over existing numbers. The ideal population, for calculation purposes, engages in all mating combinations, including self-fertilization. Most species have obligatory cross-fertilization, which slightly reduces the potential of inbreeding. Also, if multiple matings occur together with sperm storage, the genetic contribution of males can be higher than actual male numbers during short declines in density. If selection acts differently on the sexes or if drift affects only one sex, then gene frequencies may well differ between the randomly mating sexes. Robertson (1965) has drawn attention to the consequence, i.e., the resulting frequency of heterozygotes is increased. If p_m, q_m and p_f, q_f are gene frequencies of males and females, respectively, then the observed proportion of heterozygotes, H^0, is $p_m q_f + p_f q_m$, while the expected proportion, H^e, based on an equilibrium assumption, is:

$$H^e = 2 \left(\frac{p_m + p_f}{2} \right) \left(1 - \frac{p_m + p_f}{2} \right).$$

Substituting any two sets of p and q values into the expressions reveals $H^0 > H^e$, which is occasionally identified as the Robertson Effect. This property is another way that the move toward homozygosity by drift may be impeded.

Fisher (1930a) suggested that a basic question concerns the magnitude of selection necessary for it to be effective. This magnitude can be judged as a function of the population size exposed to the selection. If a suboptimal type's before-selection number is 1,000, and selection removes one individual, then $w = 0.999$ and $s = 0.001$ or the reciprocal of the before-selection numbers. If selection were less than 0.001, it would be ineffective. A population consisting of 100 individuals would not lose any individuals by s values less than 0.01. Selection coefficients that are effective in the larger group are ineffective in the smaller one. The selective loss envisioned in this way applies to mortality or inability to breed within one generation. Wright has considered the magnitude of s necessary to offset the effect of drift. Drift and inbreeding both lead to frequencies of homozygous classes higher than Hardy-Weinberg expectations, and these departures from expectation may be measured by the inbreeding coefficient of the population. This coefficient can be expressed also as a function of population numbers. If a population consists of 100 diploid individuals and if no previous inbreeding or drift has occurred, then genes of identical descent would not exist at any locus. For example, for the a locus, 200 alleles unrelated through a common ancestor would exist, such as $a^1, a^2, \ldots a^{200}$. For simplicity, assume that all alleles can unite, e.g., male and female gametes arise in the same individual. For this locus, 200 gamete types exist. After selecting any one type, the probability of selecting a second gamete identical to the first choice is $1/200$ or $1/2N$. (Actually, the value is slightly less, since one individual usually lacks the ability to fertilize its own gametes and sampling was without replacement. The difference for most values of N is negligible.) Since the coefficient of inbreeding is the probability that two genes in a zygote are identical, i.e., alike in structure and from a common ancestor, the $1/2N$ value is the coefficient for a single generation of random combination of gametes in a population of N individuals. Wright notes that if selection is to offset drift, then $s > 1/2N$. Fisher's view on the magnitude of an effective s can be expressed as $s > 1/N$. The difference lies in the fact that Wright's usage allows s to be determined by both survival and reproductive differences, and, rather than measuring the number of affected individuals, it measures the disadvantage that homozygous classes must experience.

Drift is often associated, without qualification, to small population numbers; however, the expression $s < 1/2N$, for identifying conditions necessary for drift, reveals the necessity of qualification. One set of individual numbers may involve drift for one locus and not for another if s values are independent for the loci. Thus, if N is 100, and s for locus A and locus B is 0.009 and 0.003, respectively, then even if p for locus $A = p$ for locus B, the observation that $0.003 < 1/200$ and $0.009 > 1/200$ shows that drift may affect locus B alone. Since most populations are well over 100 and most estimates of selection $\geqslant 0.01$, the likelihood of drift seems low if the selection estimates are even roughly accurate and representative.

Actually, if the increase of homozygous frequencies produced by the coefficient $1/2N$ is not large enough to result in at least one homozygous individual above the number expected with no drift, then the process is a statistical abstraction. Accordingly, small populations do constitute the setting where actual drift may be expected.

Some natural populations do fall to low numbers. The Whooping Crane is a well-documented example, but the frequency of such restriction in numbers not also being followed by extinction is unknown. In the formative days of population genetics, the leading contributors envisioned most s values as small, e.g., 0.01, 0.001, etc., and thus as difficult to verify, but as having enough time to explain evolution. We now know that all s values are not so small, and with larger s values, drift must be greater in intensity, or effective populations smaller, for chance to guide changes in p and q. Also, if environmental conditions change, bringing a decline in numbers, then a major increase in selection seems to have developed. Another line of evidence concerns recent studies using electrophoresis of proteins to indirectly measure heterozygosity at loci sensitive to the method. Such studies, cited below in other contexts, reveal more heterozygosity than could be expected if drift had been operative in the populations' recent history. Nonetheless, genetic drift as an alternative to selection in explaining changes in gene frequencies is still an attractive possibility to many.

Another source of p and q change is occasionally described as meiotic drive, preferential segregation, segregation distortion, etc. These terms simply mean gametic selection.

TERMINOLOGY

Three patterns exist for response to selection, depending on the position and number of optimal phenotypes within a trait's frequency distribution. These types of selection are illustrated in Figure 3. Basically, stabilizing selection recognizes the distribution's mean as the optimal class, with variation reduced above and below this value. Directional selection favors an optimal class existing toward one of the two distribution tails and producing a skewed after-selection distribution. Occasionally, such selection is described as linear directional selection, with fitness increasing toward an extreme value in the observed distribution. These terms are generally used for continuous variation determined by many gene pairs. Where selection operates on genotypes of a single locus and the heterozygote is not favored, the situation is similar to directional selection, and one allele moves toward fixation. In balanced polymorphism, on the other

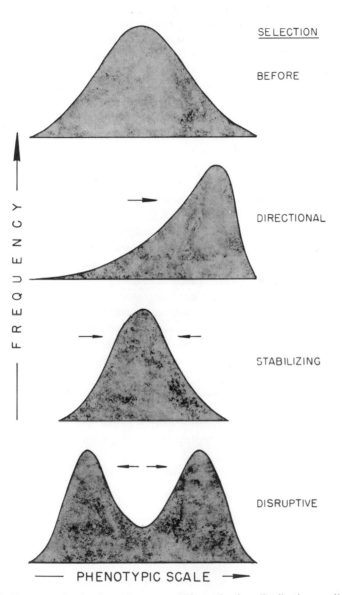

Figure 3. Patterns of selection. The upper before-selection distribution applies to each of the three lower after-selection curves. Arrows indicate the direction selection moves the before-selection mode.

hand, selection favors the heterozygote, often envisioned as the inter-mediate type, but this system is not equivalent to stabilizing selection. Balanced polymorphism does not move toward homozygosity, while sta-bilizing selection does, as described below. Disruptive selection envisions more than one optimal class, though not equally fit and not including the before-selection mean. The after-selection distribution becomes bi- to multi-modal. These terms have come into fairly standard usage; however, similar or nearly similar concepts are in the literature under different names. Simpson (1944) introduced the names "centripetal" and "centrifu-gal" selection, which are largely synonymous to stabilizing and directional selection, respectively. Waddington (1957) has used the terms "normalis-ing" and "canalising" selection, which are similar to the stabilizing concept but involve other features, such as several genotypes being channeled into a similar phenotypic expression. The term "dispersive selection" has been used in the disruptive sense by Dobzhansky (1970), and Mayr (1974), and other terms exist relating to other processes. For instance, where homo-zygotes and heterozygotes are favored, the terms homo- and hetero-selection may be used, and cyclical selection envisions s to vary from plus to minus values with changes such as season effects. Reciprocal selection recognizes a feedback effect between a parasite and its host, and ecologists have introduced terms such as r and k selection. The use of r selection implies that individuals with high intrinsic rates of increase are optimal, while k selection suggests that competitive ability for limited resources is the overriding influence. Other concepts with associated terminologies are discussed in more length below in relation to varying s values and units of selection.

The concept described above as fitness has been discussed under the terms "selection value," "adaptive value," "survival value," and "selective advantage." Unfortunately, all of these terms, including fitness, have been used with both rigorous and loose meaning. For instance, "survival value" may appear as the specific function of some morphological structure and as an algebraic factor in the same article. Carson (1961) uses the concept in a different way when he observes that population size, when environ-mental conditions are similar, is a measure of reproductive success. Thus he considers population numbers as relative population fitness. Fitness has, on occasion, been recognized as a special case for man. Namely, the family with a large number of children to support faces worry and suffering not experienced by less fecund parents, and the more fecund family fails to recognize their "blessing." Various religious and racial groups may also take exception to other such groups that are characterized by higher reproductive rates. A number of social and economic problems arise, and some authorities state that it is impossible to equate high reproductive rates with high fitness in man; they stress that the optimal rate of reproduction is not the highest possible one. The notion that population size may be regulated by selection for a lower reproductive rate has been identified with group selection discussed below. At this point, however, it seems valid to observe that fitness, in the evolutionary sense, measures an individual's genetic representation in subsequent generations. Fitness does not concern one's peace of mind or a comfortable bank account beyond the extent that such attributes are required in raising offspring to the mating age. The advantages of moderate population numbers and minimal requirements of child support are clear, but these conditions do not reflect high genetic fitness and cannot be selected for in any known way.

The selection coefficient has also been described as selection pressure or intensity. The latter term may imply the mean s value when specific genotypes are not recognized. This usage is followed in a later chapter; however, selection intensity has also been used in other ways. One form gives intensity as the number of standard deviations a group of selected individuals are removed from a population's mean before selection, also named the standardized differential. Still another usage traces to the work of Gause (1934) where he gives selection intensity as the unrealized part of potential increase in numbers divided by the realized increase in numbers. Numerous other terms exist that are largely associated with breeding programs in applied biology and that have little direct bearing on the evolutionary interest in natural selection.

Chapter 3

Evidence
for Selection

Evolutionary literature abounds with studies purporting to demonstrate selection. The majority of such conclusions are probably valid, but individual studies on natural populations often rely heavily on unverified assumptions, and the diverse nature of observations offered as evidence largely constitutes the case for selection. The first body of information consists of associations found between characters and habitat attributes conveying increased likelihood of survival. Presumably, any attribute or behavior enhancing survival and/or reproduction can be produced as evidence; however, most studies involve survival relative to predation. These investigations concern cryptic, mimetic, and warning color patterns or comparable structures. Clines and associated phenomena are also associated with inferred survival differences; however, many clines exist without an obvious relation to survival, and the negative argument is often invoked. This reasoning attempts to show that other alternatives do not explain character associations and selection wins by exclusion. A second line of evidence concerns advantageous expressions controlled by the background genotype. A more rigorous documentation comes from statistically significant differences between observed and expected character frequencies. The expected values derive from properties of the specific genetic mechanism, mating system, and equal likelihood of survival and reproduction. The environmental selection agent producing this kind of evidence frequently remains completely unknown.

CRYPTIC, MIMETIC, AND WARNING PATTERNS

Cott (1940), Grant (1963), Mayr (1963), Ford (1975), Fogden and Fogden (1974) and others thoroughly discuss and identify examples of these
patterns, and the inherent assumptions of such evidence are mainly examined here. The near-perfect resemblance of a sphingid lepidopterous larva
to a curled and battered leaf, the thorn-like protonum of membracid
treehoppers projecting smoothly out from the matching bark, the phas-
matid walking-sticks dangling at appropriate angles on their shrubby habitats, and hundreds of other similar cases give a convincing case for
improving protection from predation. Other patterns involve advertising
one's presence but bearing strong resemblance to some unpleasant component of the environment. For instance, the mimics of stinging wasps, ants,
and even of snakes by the homopteran, *Laternaria*. These examples are
assumed to reflect selection responses of prey in systems where predators
search by using visual images, much as we do. Similar responses of
predators can be cited where their cryptic appearance allows undetected
approach to prey or provides concealment for a subsequent ambush, or, by
mimicry, lures a prey to its demise. Biologists have long presented these
associations as proof of what selection might do, and the argument is no less
sound today than during Darwin's era. Ecologists have assembled a large
body of concept and data on predator-prey systems, though most are not
directly related to our quest for selection evidence. The following relationships bear more directly on the natural selection aspect.

Many of the crypsis examples reflect selection evidence for only a
small percentage of an individual's life. Many elaborate adaptations often
cited occur, for instance, only in the adult stage, yet this period is, for a
large number of prey species, typically the shortest interval of the life
cycle. This observation relates to the magnitude of selection in the following way: The term *selection effect*, discussed below, is given as the
proportion of total loss due to selection. For a given interval of time, a
small effect may result from low selection coefficients and/or small frequencies of suboptimal phenotypes. A large effect, on the other hand,
associates with a reasonably high *s* value. If the selection effect is small, for
whatever cause, random fluctuations of mortality may frequently swamp
out its expression. A small selection effect will also predictably evolve
less-perfected development of an advantage-giving trait than would occur
under a large selection coefficient. As the interval a given selection coefficient operates becomes shorter, its selection effect will also decrease.
Consequently, when we observe well-perfected development of an advantage-giving attribute that exists for only a short period of the individual's

life, we may conclude that rather high s values have operated to generate the adaptive expression. Time presumably does not fully compensate for small selection effects since one can clearly recognize remarkably advantageous traits as opposed to less perfected ones. Blest (1963) further contends that selection may actually operate to shorten the period when a cryptic costume is worn. He envisions that conspecific neighbors of cryptic species constitute a danger. When a predator locates one individual, he will presumably find the second more easily and soon not be deceived by the crypsis. If cryptic individuals die quickly after reproduction, predators are less likely to learn their secret, and an altruistic event occurs. The genetical model compatible with Blest's suggestion is given by Hamilton (1964a, b) and discussed in a later chapter.

Adult life expectancies are not always short. If exposure to selection is directly proportional to adult life spans, then long-lived prey species will be affected more adversely by a given predation rate than short-lived prey. Long-lived prey should, therefore, possess a proportionately higher reproductive output per individual or express more elaborate individual adaptations toward lowering predation. The prediction assumes predation rates are rather similar for short- and long-lived prey. Data on these questions are meager, but the latter solution seems to have some support in mimicry systems. Mimicry is particularly well developed in tropical butterflies, and it is noteworthy that Benson and Emmel (1973) and Erhlich and Gilbert (1973) recently report much longer adult life spans for such butterflies. Less variable predation rates may also exist in tropical habitats. Again, data are scanty, but such rates in temperate latitudes were found to be variable in snails and butterflies (Cain and Currey, 1968; Shapiro, 1974). Life spans, by insect standards, are also rather long in some warningly colored species, e.g., wasps and bees. In this context Blest's concept views the neighbor as an asset, since a predator is less likely to remain uneducated if his opportunity for learning is prolonged. Again, altruism may be invoked with this explanation.

The theory of mimicry is a classic in the development of evolutionary concept and substantial data support much of the theory. Ford (1975) reviews this material in a very convincing fashion. Nonetheless, some assumptions rest on rather little data and the most logical of several possible explanations is not clear for some stages of the theory.

Batesian mimicry involves a distasteful or unpleasant species, the model, and one or more eatable species, the mimics. Predators are believed to sample food somewhat randomly and, by learning, develop a visual-image search method for suitable prey. The requirement of learning ability leads to the assumption that the predators consist largely of vertebrates;

however, a few cases, at least, exist of vertebrate mimics and models. The crested rat, *Lophiomys ibeanus*, mimics the African zorilla or polecat, *Ictonyx striatus*, the value consisting of the respect predators pay to the latter's unpleasant odor. Black African birds or drongos, genus *Dicrurus*, are apparently distasteful and also possess an offensive smell, with the result that they appear to have been mimicked by a flycatcher, a shrike, and a tit (Fogden and Fogden, 1974). The cause of unpleasantness in a prey, such as distasteful or poisonous attributes, may be acquired adventitiously. An example is the familiar monarch butterfly, *Danaus plexippus*, which feeds in the caterpillar stage on milkweed plants. These plants vary geographically in toxic qualities, and individuals feeding on toxic milkweeds accumulate the chemicals in their bodies and are quite distasteful to predators. Individuals feeding on nontoxic plants are said to practice *automimicry*.

The unpleasant items are soon identified by predators and subsequently avoided. Mimics of such models enjoy the avoidance afforded the model, but theoretically have been assumed to be less numerous. If the learning predators frequently encounter the edible mimics, the image would be associated with suitable food too often for avoidance behavior to develop. The theory thus envisions the mimic as committing its population to a lower density than the model when entering this line of evolution. *Mullerian mimicry* consists of two to several species with unpleasant attributes sharing a common phenotype. The predators thus learn at a more rapid rate by receiving lessons more frequently. Presumably, members of a Mullerian system would not require the density adjustment noted for the Batesian system. In actual practice, these systems can be complex, with one species being part of a Mullerian system in one area of its range and part of a Batesian system elsewhere. Also, if food became scarce and the unpleasantness of Mullerian models is unequal, the system could phase into a Batesian operation, with predators recognizing the least offensive member of the system as poor but usable food. Distinguishing between systems is not a straightforward process.

The theory requires that the predator try an item as food and accept or reject it as the matter dictates. The innate ability to learn allows a predator to develop numerous and appropriate responses to different stimuli coming from different potential prey. Presumably, an instinctive reaction to each prey stimulus would otherwise entail several independent genetic determiners and be less flexible. The cryptic or warning color patterns of prey are, however, associated with instinctive behaviors putting the patterns to best advantage. In this case, one instinctive behavior gives a common reaction suitable for different predators, e.g., concealment or escape. The experience a predator encounters in sampling potential food can't be lethal, for no predators with acquired education would accumulate. This possibility may be uncommon; however, coral snakes and black widow spiders have been suggested as models (Levi, 1965). Such models may involve only a thin line between unpleasantness and lethality, but the latter is inconsistent with the theory in which predators learn. Biologists are occasionally hesitant to accept explanations involving instinctive fear-like avoidance, yet readily invoke instinctive behavior for the stereotyped releaser-response steps of courtship, etc. An instinctive avoidance of unpleasantness verging on lethality requires no greater assumption than is required in the ability to learn. Where this possibility has been tested, an innate avoidance has been documented for novel prey (Coppinger 1969, 1970). An instinctive avoidance by a predator would certainly not inhibit mimicry of the model in question. A possible explanation avoiding instinctive behavior by predators relative to very poisonous snakes suggests that the deadly snakes mimic less-deadly ones and the predators learn by way of the latter. The term *Mertensian mimicry* is applied to the system. A number of facts do not support this scheme. First, the usually deadly coral snakes (presumably the mimics) and the supposed models do not coincide well in geographic range. Secondly, experienced field collectors often observe that the supposed models seem more secretive and scarce than the deadly mimics, an unlikely feature of Batesian mimicry.

If the predator, while taking the model and learning of his mistake, damages the individual to the point that it dies or is impaired and experiences a reduced reproductive output, then the value of its unpleasantness is questionable. The frequency of such damage in all contacts with models bears on the question; however, the concept of the individual as the unit of selection is at stake. This question troubled Fisher in 1930 (1958a, p. 177), and he came rather close to developing a process currently termed *kin selection.* Carpenter (1941) compared the frequency of beak-marks on butterflies having different levels of acceptability by birds. He generally concluded that the less acceptable species possessed more marks, indicating that they had successfully survived a predator contact. Presumably, the fewer marks on edible species indicated that the birds succeeded in eating most of those that were caught. The data on this point are not extensive but suggest that models are hardy creatures. Edmunds (1974) has recently reviewed beak mark studies, and he concluded that several sources of possible error exist that have not received sufficient attention when data are interpreted. Fogden and Fogden (1974) cite several cases where models are seen to survive a predator's attack, and note that poisons are perhaps combined with an emetic. This feature may save the predator, now endowed with a memory to avoid a specific model, but does not imply that the model has survived. Mimics in Batesian systems are also typically smaller than their models. Birds appear to take larger insects as food when given a choice and a survival benefit may accrue to mimics by small body size. This notion assumes the color and pattern concern a predator's memory more than size. The evidence for these possibilities are given by Ford (1975).

The density factor claimed for Batesian mimicry is understandable in that most predator contacts involve the model thus keeping their senses keen to the unpleasantness. Evolution of the numerical adjustment between model and mimic is less clear. Essentially, the reproductive output of the mimic must be lower than that of the model. Development by natural selection of a decreasing reproductive output is incompatible with the concept of selection favoring fitness of individuals. Haldane (1953) did present a mechanism where the spread of an advantageous gene could reduce population size. His example involved a moth attacked at one stage of life by a parasite and at another by a predator (vision-oriented bird). This scheme has no obvious relation to mimicry without adding some unlikely assumptions. Are mimics possibly restricted to species that independently have their numbers limited to values below their model? Theoretically, this dilemma can be avoided in other ways. If the degree of a model's unpleasantness is high and the predator has a well-developed capacity for memory, then fewer model encounters may develop avoidance behavior by the predator. Brower (1960) has suggested this possibility from experimental work with artificial prey and birds. The extent of this mechanism in actual systems is unknown, but other possibilities exist. The mimics in Batesian systems often occur only in females. The males' color and pattern are thought to function in species recognition, an advantage outweighing the reduced predation conferred by mimicry. Only female numbers relate, therefore, to the model. Another observation is that polymorphism occurs more commonly in Batesian systems (Ford, 1975). Namely, a species evolving mimicry may, by polymorphism, partition its numbers over several models and again avoid the need to reduce density. A less frequently suggested relationship for density concerns warningly colored species (Fogden and Fogden, 1974). In this concept, densities of warningly colored species must remain relatively low for the protective attribute to remain effective. Otherwise, their high numbers would presumably present such a ripe source of food that some predator would evolve the ability to cope with their defense. If one accepts the premise that high numbers is a liability in this way, the difficulty remains in explaining how selection has limited their population growth. The suggestion is also troublesome for Batesian mimicry, where the unpleasant model occurs in high densities.

Protection through mimetic systems is said to involve a continual change, in comparison to the cryptic route of escape. A moth, mantid, etc. resting on the bark of a tree (or rock) and bearing an uncanny resemblance to its background hardly influences the tree's fitness. A mimic in the Batesian concept is credited with playing a different role. When a young and learning, or old but forgetful, predator takes a mimic, it presumably obtains a suitable bit of food and can be expected to try a repeat performance at the next opportunity. That encounter may, however, be with the model. If the model suffers, its fitness will decline and selection is supposed to operate toward separating the model phenotype from that shared by the troublesome mimic. Of course, selection would presumably also make appropriate adjustments in the mimic. Brower and Brower (1972) discuss this phenomenon as *advergent evolution.* Again, the outcome of predator-model contact is critical. The predator is credited here with lowering fitness of the model because of confusion with a numerous and closely matched mimic. Yet, in the initial steps of the theory, the predator was assumed not to be lowering the model's fitness, thus conferring an advantage on its being unpleasant. This portion of the theory appears difficult because a direct advantage to the individual is sought. If kin selection is allowed, model mortality at any time poses no contradictions.

Other mimicry systems have been described, and in one intriguing scheme the prey is the model and the predator is the mimic. The relationship presumably allows the predator to approach the prey or lure the prey into a position to be attacked. The system is termed *aggressive mimicry* and the predators are less likely to be vertebrates than in the above examples. The case with lampyrid fireflies is an example (Lloyd, 1965). The female of one species, the predator, mimics the flash pattern of males in a second species. The responding females of the prey are thus greeted by a predator rather than a mate. Two other cases indicate the range of interactions such mimicry may assume. In Borneo, a crab spider, *Misumena vatia*, takes advantage of butterflies that alight on fresh bird droppings, apparently attracted to the salts, etc., available. The spider mimics the droppings in color and spins a small web about its body to further enhance the effect; it then awaits the arrival of deceived butterflies. A Malaysian mantid, *Hymenopus coronatus*, closely mimics a flower and positions itself near similarly colored true flowers, where unwary insects fall prey as they move from flower to "flower." Predatory robberflies mimic some insects, presumably to enhance their unsuspected approach as they either capture a prey or place eggs in a position that allows the larvae to feed on the prey's larvae. Brower, Brower, and Westcott (1960) studied this system with robberflies and bumblebees. All such systems are not well understood. Robberflies of the genus *Hyperechia* closely mimic carpenter-bees, *Xylocopa*, in India and Africa. The resemblance is so close that each species of *Hyperechia* matches a particular species of *Xylocopa*. The larvae of the two forms live in similar areas but no convincing data otherwise connect the two types. The robberflies occasionally take carpenter-bees as food, although they do not specialize on the bees which they resemble and take many other insects, too. Oldroyd (1964) concludes that this point is "very inconvenient for mimetic theory." As noted above, an elaborate selection response may occur for a small percentage of an individual's life span, but a well-developed mimicry seems unlikely to evolve for obtaining only a small percentage of one's food. Oldroyd's question suggests that aggressive mimicry may not be the best explanation. The robberflies may be simply envisioned as Batesian mimics of bees; but if so, why mimic specific species in such detail? Predators are unlikely to know the local bees to that degree.

The basic assumption in cryptic systems is an increase in likelihood of survival when the cryptic pattern and background are properly associated. For many years only a few observations supported this assumption. Poulton (1887) seems to have initiated the experimental approach to verification. di Cesnola (1904) and Gerould (1921) reflect subsequent development of experimental proofs. Various prey, such as mantids and butterflies, were used and birds served as predators. Collectively, their efforts document that predators possess the potential for discrimination, in that the cryptic patterns with mismatched backgrounds were more vulnerable. Popham (1941) summarizes much of the experimental evidence of the survival value of color patterns available by 1940. Recent workers have refined methods for analyzing numerous conditions, and these techniques are discussed by Manly et al. (1972). One basic assumption in mimicry has largely been verified by Brower (1958a, b, c). Her studies asked the same question as Carpenter's report; i.e., do predators have the ability to learn avoidance of unpleasant items for food? Data for the instinctive means of avoidance are few; however, Coppinger's studies mentioned above clearly testify that such attributes exist and best explain the evolution of warning coloration for many species.

The cryptic, mimetic, and warning costumes of animals, developed by natural selection, still appear as intriguing as they did on their initial recognition, and the unanswered questions stand as challenging areas of study.

SEXUAL SELECTION

Another early source for selection evidence was advanced under Darwin's term "sexual selection." Darwin (1859) distinguishes the process as ". . . this form of selection depends, not on a struggle for existence, but on a struggle between the individuals of one sex, generally the males, for the possession of the other sex. The result is not death to the unsuccessful competitor, but few or no offspring." Darwin provides several examples that he felt were explained in this way, such as antlers on stags, large mandibles on male stag beetles, colorful plumage of many male birds, etc. His definition does not specifically require female choice but his examples and discussion often imply it.

Fisher (1930a, 1958a) drew attention to the fact that two components of sexual selection may exist. First, some attributes in sexual selection become advantageous in male interactions where fighting, pseudofighting, displays, etc., are addressed to other males in order to inhibit the attentions of the submissive male toward females. An improved expression of such characters will, within limits, increase the reproductive prospects of the dominant males. Presumably, a difference in sexual vigor between male types, not involving intermale contact, could achieve the same effect for more active males. Secondly, male attributes, such as, in male birds, colorful plumage that stimulates females to mate, should also increase in frequency and degree of expression. Female choice is required for sexual selection in the second component. Intermale contacts may identify a dominant male in the absence of females. Where females are present at the contacts, they may actively choose the dominant male, though this process seems not required. A distinction between female choice and submissive male inhibition may frequently be difficult. Both forms of selection can lead to sexually dimorphic characters; however, many such characters are more logically explained as survival attributes by conventional natural selection.

Mayr (1972) concludes that the observations becoming available since Darwin's era require a modification of the sexual selection concept. First, he notes that male fighting, displays, etc. are largely territorial where mating and nesting sites, etc. are being established. Frequently, the females are absent during these contests, and when they arrive to mate the habitats acceptable for these activities are occupied only by males already successful in territorial behavior. The females seem to have rather little chance to choose between territorial abilities. Mayr's remarks relate largely to birds and other vertebrates, but the same events seem true for other groups, particularly for dragonflies (Johnson, 1964). If the criterion of female choice is adopted for sexual selection, then these male characters evolve by differential reproductive success or natural selection. This view is more narrow than envisioned by Darwin and also removes differences in sexual vigor from sexual selection.

Ample evidence does exist that females, for most species, are the more discriminating sex in accepting a mate. They may exercise choice only between males that maintain themselves successfully on the breeding sites. Where sex ratios are near equality, a rather common observation, and females practice discrimination, mating systems may become nonrandom. The extent of this possibility in nature seems to have received little attention, but assumptions associated with random mating are probably unfounded if sexual selection is known to exist.

A particularly well-documented case of sexual selection exists in the scorpionfly, *Bittacus apicalis* (Thornhill, personal communication). This species practices nuptial feeding, whereby the male provides the female with a meal (some small arthropod) during the copulation period. Males obtain a prey and then position themselves by hanging in the vegetation and releasing pheromones, thus advertising for a mate. Once a female reaches the male, she then exercises discrimination based on size of the offered food. Below a threshold size, females will rarely mate. The advantage to the female lies in the fact that she receives sufficient nutrition to last between matings; ovarian development fails to begin when nutrient levels fall below this size. The male offering prey that is too small mates infrequently and, in addition, may lose the effect of the few matings achieved. The latter consequence develops from the fact that females having received an insufficient food supply proceed, without oviposition, to mate again, perhaps because of the necessity for the required food. The last copulation probably results in displacing sperm of the first mating, and the subsequent eggs are fertilized in large part by the last mating. Thornhill's observations reveal high levels of female discrimination and yet males offering insufficient prey continue to exist. In fact, a single male may offer both suitable and unsuitable food size on two different occasions. The variability of male choice in food size perhaps has a large environmental component. Otherwise, the offering of small prey is a practice that would be expected in far lower frequencies than observed.

Trivers (1972) approaches the explanation in another way by noting that some activities of individuals improve survival of their young rather than increase their chances for more matings or for a large number of young. Such attributes are termed *parental investment.* He further notes that this investment is generally not equal between the sexes. When the investment is unequal, the sex investing least (usually males) reacts to the other sex as a resource and engages in intrasexual competition. In this way, Trivers explains the common observation of female choice (the high investment parent) and of the setting most likely to involve sexual selection. Characters initially evolved by natural selection may subsequently come to be favored by sexual selection, and the historical origin of a trait may not be evident from the current nature of its advantage.

The literature occasionally blurs concepts when attention is focused on a mechanism known as sexual or ethological isolation. The latter terms describe one of several processes, known as isolation mechanisms, by which gene pools of separate species are retained intact against possible hybridization. This specific isolation mechanism concerns behavioral attributes insuring that matings involve conspecific individuals. The clues for recognizing a conspecific may exist in: a complex courtship, vocal stimuli, minor differences in structural appendages used in pairing, etc. Since the same structure or attribute may well be sensitive to both sexual selection and sexual isolation, it is not surprising that the original meaning of the former process is occasionally indistinct. Nonetheless, two different processes exist. With sexual selection, the reduced numbers of offspring result from lower competitive ability or attractive value to the opposite sex. One sex, usually the female, chooses within varieties of conspecific members of the opposite sex. The overall effect is a selection for attributes within one sex that enhance mating acceptance by the opposite sex. With sexual isolation, the reduced numbers of offspring are due to zygotes having lower fitness as a consequence of interspecific parents. A choice is invariably involved, by either males or females or both, between inter- and intraspecific possibilities for mating. The overall effect is selection favoring distinct expression of conspecific features that may have little relation to mating vigour. Parsons (1967) summarizes efforts to document these processes in laboratory mating chambers as measured by coefficients of vigour and isolation. Sexual isolation has received most study consistent with the wide interest in speciation, and perhaps due to difficulty in observing mating behavior of many species. The evidence for sexual selection is nonetheless convincing, and a series of essays edited by Campbell (1972) summarizes current opinions.

CLINES AND AREA EFFECTS

The term *cline* was originally suggested by Huxley (1939) and adopted widely by students of evolution. A cline is a directional change of character or gene frequency within a species over geographic distance. A graphic plot of frequencies with distance usually reveals a rather smooth ascending or descending curve. Fewer examples involve large frequency changes over short units of distance, the so-called stepped clines. A cline concerns, therefore, the properties of a single character. Several characters may show a similar direction of change but their slopes may well differ. Other characters may change from site to site independently. Such variation is often termed *discordant* or *site specific*. Reverse clines or character displacement are occasionally used in describing an attribute in two species (Ford, 1975). A marked discontinuity often develops in the frequency curve within the area of sympatry.

Usually, clines have been documented by scoring attributes in field-collected adults. These samples provide estimates following some mean interval of possible selection and are recognized as after-selection values. If before-selection values are measured along the cline, the position and slope of the curve may differ from a curve taken from after-selection values. Recall that genotype proportions may change within a generation without changing gene frequencies. Hayne (1950) used such estimates, in part, with a study of deer mice mentioned below. He bred field-collected adults in the laboratory to obtain progeny (before-selection) for sibship comparisons. If cline data are being compared among several generations, comparable samples regarding exposure to selection are desirable.

Two concepts exist for explaining clinal patterns within a species in which fitness differences exist but develop in quite different ways. The principal force believed to be operating in addition to selection is migration; consequently, documenting a cline under these assumptions constitutes selection evidence. The generally accepted model has fitness at any point on the distance scale developing directly from the impact of that specific environment. Fitness is thus high at one end of the distance range and decreases to a low value at the opposite end. If fitness is expressed in a relative fashion, a point somewhere along the cline may represent a fitness of 1.0. Below this point fitness values become progressively <1.0 and above the point, fitness values become progressively >1.0. This intermediate point may be termed a *fitness boundary*. Each population along the cline has its gene frequencies in an equilibrium determined by the ratio of fitnesses (Fisher's Expression), which prevents a direct estimate of the selection coefficients. Haldane (1948) used the slope of the cline curve to estimate selection coefficients and his methods are discussed in a following chapter. A number of clines conform to this interpretation rather well.

A frequently mentioned example involves the deer mouse, *Peromyscus polionotus*, originally documented by Sumner (1926, 1930). The populations living on the white sandy beaches bordering the Gulf of Mexico in northwest Florida, often recognized as a subspecies, have a distinctly whitish pelage color. As one traces the populations inland, the color darkens, although at different rates for different parts of the body, presumably due to different controlling genes. The melanism of *Biston betularia*, a moth studied intensively in England by Kettlewell (1956a, b; 1958), Clarke and Sheppard (1966), and more recently by Bishop (1972), possesses a dark morph, *carbonaria*. The frequency of *carbonaria* is high in the polluted conditions of industrial cities and decreases in a clinal pattern as populations are sampled toward the rural, less polluted habitats. Such examples consist of animals exhibiting a blend of body color with background shades. Cryptic-type survival benefits appear most probable. This conclusion is largely verified by experimental data. Dice (1947) demonstrated that rodents matching their background in color enjoyed higher survival from owl predation, a natural enemy of *Peromyscus*. Kettlewell (1956a) has also shown that bird predation favors the dark *carbonaria* morph in industrialized areas. Perhaps the most direct support comes from Bishop's work (1972). He studied a cline of *Biston betularia* around Liverpool, where he demonstrated an actual gradient of selection coefficients along the cline. The use of electrophoresis has been applied to the study of geographic variation, revealing clines for enzymatic proteins correlating reasonably well with possible agents of selection. The bryozoan *Schizoporella unicornis* has an allele coding for leucine aminopeptidase that varies significantly over distances as low as 20 miles along the east coast of the United States (Schopf and Gooch, 1971). The fish *Catastomus clarkii*, a sucker, has an allele for esterase, est-I^a, that varies from 0.18 to 1.00 in a latitudinal pattern along the Colorado River drainage. The above change occurs over only 7.6 degrees of latitude (Koehn and Rasmussen, 1967). The investigators conclude that observations relate to a correlated temperature gradient. The familiar *Drosophila melanogaster*, scored by the alcohol dehydrogenase locus, exhibits clines in Asia and eastern Europe correlating with altitudes (Grossman, Koreneva, and Ulitskaya, 1969). Again, temperature is a logical causative agent though other reasonable alternatives exist. The concentration of oxygen and water turbulence correlates best to the cline of an allele for lactate dehydrogenase in the fish *Anaoplorchus purpurescens*, a rocky intertidal blenny, found near Puget Sound (Johnson, 1971). A cline correlating with latitude also exists for a hemoglobin allele in the marine bivalve *Anadara trapezia* of Australia, but the likely environmental correlate is not known (O'Gower and Nicol, 1968).

Other cline-type variations having suggestive survival functions are the so-called ecological rules. The changes of characters are similar in many species and follow similar directional trends. For instance, Bergmann's rule

associates large body size with cooler habitats in warm-blooded vertebrates, Allen's rule describes shorter body extremities in cooler habitats,
and Gloger's rule relates darker, melanic pigments to moist habitats and
light browns with xeric conditions. Different authors suggest different
advantages for these observations but all agree that survival values are
reflected.

A cline developed by selection and migration presumably remains
stable as long as these forces are unchanged. An unstable cline could
express the advancing wave of some advantageous gene moving through
the population (Fisher, 1950), a condition comparable to transient polymorphism. The likelihood of fortuitously collecting data during such a
change is low, and another complication of assuming stability may be
more common. Incomplete data may suggest a cline when the variation is
actually seasonal and involves growth rates. Body size in the damselfly,
Calopteryx dimidiata, varies seasonally; however, this change was unrecognized for many years and size was thus a criterion for a "smaller northern
species" (Johnson, 1973). The stability assumption has been verified for
relatively few clines. Hayne (1950) and Selander et al. (1971) re-examined
the *Peromyscus polionotus* cline and found somewhat different patterns of
variation than Sumner. Selander is inclined to consider population subdivision and possibly genetic drift as contributing to the variability, so that
the case is no longer an uncomplicated example. The data for *Biston
betularia* represent a recent and rapid evolutionary change associated with
development of industry (Kettlewell, 1973). An example of non-industrial
melanism occurs in the moth *Amathes glareosa* in the Shetland Islands.
The variation is distinctly clinal, and data here indicate marked stability
(Kettlewell and Berry, 1969).

Data for a number of clines exist without a correlated environmental
change likely to affect fitness. The color pattern in the wings of males and
the female dimorphism of the damselfly, *Calopteryx dimidiata*, constitute
examples (Johnson, 1973). To explain such clines by the above concept,
an unrecognized, environmental gradient change related to fitness is assumed. This solution may invoke pleiotrophic gene action, in which the
real advantage is not wing pattern, for instance, but an unrecognized, yet
associated, attribute. Such explanations are not entirely satisfying and a
second cline interpretation was developed by Clarke (1966). The concept
is built on reasonable genetic properties and involves a minimum of two
gene loci. Consider a single locus of two alleles, A and a, where the
heterozygote has a fitness advantage relative to its position on the distance
scale. Fitnesses can then be expressed as a function of distance, d, as
follows:

Genotypes	AA	Aa	aa
Fitness	$1-d$	1	$d.$

The value of d can be taken as varying over the distance scale from zero to one, and the equilibrium values of q, $\hat{q}$ increases in a similar direction so that high q and d values coincide. Note that d can be expressed as the Aa fitness minus that of AA. Recall that the equilibrium frequency of a, $\hat{q}$ equals $(b-a)/(2b-a-c)$ where a, b, and c are the fitness values of AA, Aa, and aa, respectively, in Fisher's Expression. Substituting the terms $1-d$, 1, and d into the expression gives $\hat{q} = d$. These $\hat{q}$ values are determined by environmental selection agents and migration, as discussed above.

Now consider a second locus with alleles B and b, where $B-$ acts as a modifier on the first locus, but where bb has no effect, i.e., is hypostatic with that locus. The modifying effects of $B-$ genotypes on the genotypes at the locus for A and a are x, y, and z. These terms are interaction coefficients and have no relation to a position on the distance scale. The fitness possibilities when considering both loci are:

	p^2 AA	$2pq$ Aa	q^2 aa
bb	$1-d$	1	d
$B-$	$1-d+x$	$1+y$	$d+z$

For B to affect the cline, it must spread in the population; i.e., the average fitness of $B-$ genotypes must exceed that for the bb genotypes. This condition results when:

1. $p^2 (1-d+x) + 2pq (1+y) + q^2 (d+z) > p^2 (1-d) + 2pq + q^2 d.$

Adjusting the inequality so that the right side becomes zero defines the necessary properties in x, y, and z terms. If p is expressed as $1-q$, and the right-hand term subtracted from the left-hand term, the following simplification results:

2. $\qquad q^2 (x+z-2y) + 2q (y-x) + x > 0.$

If, for practical purposes, B becomes fixed, the fitness terms for AA, Aa, and aa become $1-d+x$, $1+y$, and $d+z$, respectively. These values in Fisher's Expression give:

3. $$\hat{q} = \frac{d+y-x}{1+2y-x-z}.$$

When the B locus is not considered, the $\hat{q}$ equals d. Namely, $\hat{q}$ is then determined by conditions specific to its position on the geographic range. When the B locus is considered, the same geographic position will have the same $\hat{q}$ value only if $y = x = z$, an unlikely condition. So, interaction coefficients may influence the geographic curve of $\hat{q}$ values. Since $\hat{q}$ produced by introducing the interaction coefficients is $(d+y-x)/(1+2y-x-z)$, and the fitness difference between Aa and AA, without the B locus operating, equals d, the fitness difference with the B locus in operation gives a modified expression of the point as follows: $1+y-(1-d+x)=d+y-x$. Thus, the regression of $\hat{q}$ on distance is $1/(1+2y-x-z)$ and the value at fixation of B is then a function of the interaction coefficients rather than the fitness component produced by the immediate environment. If at least one of the interaction coefficients is negative, B does not become fixed and is limited by the value of q. For example, take the case where $x = -0.20$, $y = 0$, and $z = 0.20$. From expression 2 above, the left-hand terms are only greater than zero if $\hat{q} > 0.5$, thus over the values of d up to 0.5, $\hat{q} = d$. Above a value of 0.5 for d, the interaction coefficients produce $\hat{q}$ values given by expression 3 above. The values from 0.5 to 0.7 for d are:

d	$\hat{q}$	$\Delta\hat{q}$
0.7	0.9	
0.65	0.85	0.05
0.60	0.80	0.05
0.50	0.50	0.30

At $d = 0.5$, expression 3 is no longer valid and $\hat{q}$ falls to 0.5 for a $\Delta\hat{q}$ of 0.3 between d values of 0.5 and 0.6. Thus, a distinct step appears in the cline without a step in environmental conditions (Figure 1).

The spread of a gene such as B is unlikely if the x, y, and z terms significantly impeded response to local selection. Where an environmental gradient exists that significantly influences survival, i.e., where d values are measurable, the cline may be expected to correlate with environment. Characters having small d values are more likely to be influenced by interaction coefficients.

Clines developed by genetic drift of selectively neutral alleles are unlikely to occur over any appreciable distance, since the drift process would predictably show no consistent trend. Endler (1973) dismisses the drift explanation by noting that the process requires less migration than usually occurs. A more difficult assumption for the drift explanation lies

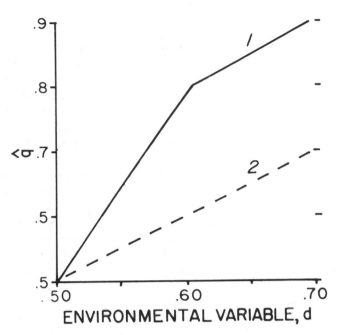

Figure 1. Clinal slopes when the equilibrium values of q, $\hat{q}$, are determined only by environmental agents along the abscissa, curve 2, and also when influenced by interaction coefficients at d values greater than 0.5, curve 1. See text for explanation.

in the notion that geographic distance can vary without involving fitness differences. No two natural habitats are entirely alike and fitness usually shows some response where studies have been made.

Contrasting to clines, one phenotypic pattern may occur over large regions involving local environmental differences. These patterns may be more common than clines. Biologists are more intrigued with variation than a lack of it, and have less frequently described the range of monotonous phenotypes. This phenotypic stability received little attention in terms of selection prior to Cain and Currey's (1963) study of *Cepaea nemoralis*. These authors used the term *area effects* to describe the phenomenon. The situation in *Cepaea* forced its consideration by biologists, since other studies in other areas found the species exhibiting correlation with habitat attributes, and was, in a fashion, explained by predatory selection. They suggest that subtle fitness differences exist, and although no obvious environmental change exists, the differences are high at the boundary of an area. Testing this notion is difficult, and Clarke (1966) suggests that areas are maintained by a co-adaptation or fitness interaction between gene pairs, as discussed above for such clines. Another snail genus, *Partula,* also expresses area effects (Murray, 1972). If internal genetic co-adaptation causes area effects, then disharmonious combinations of genes can be expected along the area's borders where individuals with one co-adaptation cross with individuals having other gene arrangements. Some evidence exists supporting this prediction in *Partula* (Murray, 1972). Interpreting area effects in this way touches upon a controversial part of evolutionary theory. The adjacent co-adapted complexes developed by Clark's cline model occur in continuous populations; however, conventional theory states that populations having different co-adaptation patterns have experienced a historical isolation. Unless rather good historical data exist for a habitat, distinguishing between these possible origins for co-adaptation may be difficult. The above discussion of geographical variation (clines) or its absence (area effects) explained observations by a genetic adjustment for high fitness. The low likelihood of selectively neutral genes drifting in a clinelike pattern was mentioned, but such genes may drift from site to site, producing a patchy distribution of allele frequencies. (Malécot, 1959; Kimura and Weiss, 1964). Such a distribution could well resemble site-specific frequencies resulting from selection, as Adamkewicz (1969) proposes for the allele giving red body color in the terrestial isopod *Armadillidium nasatum.* A small degree of migration is, however, capable of swamping out the differentiation arising from drift of neutral genes (Lewontin, 1974). If measurable migration between such sites can be documented in the field, it is presumably enough to invalidate an explanation based on neutral genes, drift, and patchy allele distribution. On the other hand, the low migration capable of destroying neutral differentiation from site to site leads to homogeneity of allele frequencies, a situation comparable to an area effect. Distinguishing between homogeneity maintained by 1) area effect selection, or 2) migration between drifting alleles, seems best pursued by examining the direct effects of the character in question on the actual components of fitness. Differentiation of populations by drift of neutral alleles is so easily obscured by small migrational input that similar allele frequencies in different populations cannot be used as selection evidence. Migration does not occur, however, between distinct species, but similar alleles may coexist in related species. If two species have been separated for a sufficient number of generations, then allele frequencies in the two species can be assumed to be independent. Lewontin (1974) reviews the data available on this comparison. The most compelling selection evidence concerns the two species *Drosophila pseudo-obscura* and *D. persimilis.* Thirty-nine loci were scored, and only four, or

10.2 percent, of the loci were judged distinctly different, the so-called diagnostic loci. Even where multiple alleles occurred, the same allele occurred in highest frequency for both species. The species had been isolated for more than a million generations.

Character frequencies may vary with time as well as distance; in fact, this observation was the theme of Darwin's case for evolution, transformation through time. Directly observing the origin of species in this way is unlikely, but single character frequencies vary more rapidly and involve expectations discussed below. Consider the dramatic abilities of microbes to develop resistant strains to antibiotics and of insects to pesticides. For intuitive but compelling evidence, such observations give a convincing case for natural selection. When frequencies change with time, values at time 0 become the expected values at time 1 with an absence of selection; thus, unlike spatial differences in gene frequencies, these temporal differences can yield estimates of selection coefficients. Since genetic polymorphism involves a lack of change in gene frequencies with time, it is worthwhile to note that an argument for selection may be built on frequency change or lack of change measured by distance or time. Thus, selection may explain all, and, in the view of some, thereby explain nothing. A more rigorous search for direct evidence is required.

STUDIES FROM NATURAL POPULATIONS

The following examples illustrate efforts to document selection operating in natural populations. In some cases, all data have originated directly from populations in the wild, while others combine both field and laboratory data in their analyses. The following properties of the selection process are essential in judging these studies.

Selection acts as a nonrandom, discriminating process and frequencies of affected characters should significantly depart from values predicted by random events. Relative survival may be measured by a minimum of two randomly collected samples within a generation, e.g., release and recapture, young and old, or living and predated, where frequencies in the former samples become expected values for the latter samples. Refinements of this approach compare mean life expectancies obtained from life table data. Such samples represent before- and after-selection periods within one generation; consequently, the nature of the mating system has no influence. Also, one need not know the actual inheritance mechanism of the traits scored. Relative reproductive success taken from life table data involves the same rationale. Investigators often assume equality for the reproductive component of fitness, however, to expedite experimental design and calculations. An inequality in reproductive output does not invalidate selection found operating on survival but may well change the net fitness.

Data overlapping two generations are best evaluated with a knowledge of inheritance for the character(s) concerned. In addition to sample collections, randomness of the mating system constitutes the basis for the expected genotype or phenotype values, and the most logical cause for departure from these values is differential survival or reproduction, thus selection. The random mating assumption may be invalidated by small population size, inbreeding, assortative mating, and migration. The likelihood of the last process affecting a specific population is small during most short-time observations. The first two possibilities are related phenomena and must be judged specifically for each study. Frequencies expected from random mating can be applied, with proper assumptions, to the analysis of both single and multiple samples. Recall the potential for misinterpreting the genotype having maximal fitness when considering a single sample (discussed in Chapter Two). In all methods, a clear distinction between before- and after-selection samples is necessary. Also, recognition of the period that a method tests for selection allows a rough prediction of selection magnitude likely to be involved. Recall that a before-selection sample scored for a single locus contains no information on selection, since the preceding mating interval restores equilibrium. Samples representing an after-selection period are required, and sufficient genotypes must be recognizable to allow an estimate of gene frequencies. From these frequencies, the expected genotype or phenotype numbers are obtained for comparison with comparable observed numbers. This approach is perhaps the most common.

An example exists with Blair's (1947) study of the buff, G, and grey, g, genes in *Peromyscus maniculatus blandus* living on soils of variable color in New Mexico. The terrain involves color change of the soil over short distances and the genes also occur in different frequencies. For example, the frequency of G was 0.567 and 0.248 at Tularosa and Alamogordo, respectively. The buff gene is dominant, and only a part of the dominant phenotypes in the live-caught samples were identified to genotype by test crosses. The estimate of gene frequencies utilized Cotterman's technique described earlier. The gene frequencies and proportion of dominants tested thus provide expected numbers in the four scored classes assuming equilibria. Blair's four major samples gave the following comparisons:

Adult Classes

Location	Buff Untested	Buff GG	Buff Gg	Grey gg
Tularosa				
Obs.	52	14	22	20
Exp.	51.9	14.2	21.7	20.2
Alamogordo				
Obs.	32	4	43	97
Exp.	31.1	6.5	39.1	99.3
Salinas				
Obs.	37	5	10	13
Exp.	36.7	5.6	9.3	13.4
Lone Butte				
Obs.	7	4	25	21
Exp.	6.6	6.1	21.3	23.0

Chi-square tests (having one degree of freedom as expected classes required q, sample size, and proportion of dominants) show no significant departures from the equilibrium hypothesis. These data are comparisons of single, after-selection samples scored for one gene locus over a portion of one generation. Any departure from equilibrium would reflect selection in one generation and small selection coefficients would pass here for neutral equilibria. The differences between samples certainly suggest response to selection over an unknown number of generations and are comparable to associations with distance discussed above. The distribution of allele frequencies has a patchy pattern, as is expected for neutral alleles in populations having very little migrational exchange. The phenotypes associated with the alleles show a reasonable correlation to the substrate color, suggesting an adaptive value in reducing predation. Also, the likelihood of migration among these mobile rodents strongly reduces the plausibility for an explanation without selection.

The much-quoted study by Fisher (1939) was an early attempt both to demonstrate and to measure selection. Fisher's material consisted of several large samples, mostly from Texas, of the locus *Paratettix cucullatus* Burmeister (formerly *P. texanus*). The species is polymorphic for color pattern involving at least 13 linked genes. He used, however, a single locus model and employed phenotypic proportions in the analysis. His method has the following rationale. If we have a two allele system, A and a with frequencies of p and q, and a sample such as:

$$\begin{array}{ccccc} \text{Genotypes} & AA & Aa & aa & N \\ \text{Numbers} & 80 & 240 & 180 & 500, \end{array}$$

the following relationships are seen. The estimate of p^2 is 80/500, giving $\sqrt{80} = p\sqrt{500}$; similarly, $\sqrt{180} = q\sqrt{500}$. Now, if the assumptions are valid, $\sqrt{80} + \sqrt{180} = \sqrt{500}$, since $p + q = 1.0$. Even though linked genes were involved, the double dominants were much less frequent than expected, and triple dominants were absent. The double dominant deficiency is discussed more fully later. The low frequency of double dominants, the coupling phase, allowed Fisher to assume that they were absent. If A is one dominant and any other is B, three gene combinations exist considered as "gametes," i.e., Ab, aB, and ab, while AB is the combination assumed absent. Mating combinations between these gametes follow a multiple allelic pattern, where each above gamete has a frequency of p, q, and r, respectively, with $p + q + r = 1.0$. Treatment of the following sample illustrates Fisher's process of evaluating assumptions.

Phenotypic Classes	Ab	aB	AB	ab	N
Frequencies	$p^2 + 2pr$	$q^2 + 2qr$	$2pq$	r^2	1.0
Sample Numbers	214	208	151	57	630

The recessive zygote r^2 is 57/630, giving $\sqrt{57} = r\sqrt{630}$. The combined frequency of individuals expressing A plus recessives is $(p + r)^2 = 271/630$, giving $\sqrt{271} = (p + r)\sqrt{630}$. Likewise for B and the recessives, $(q + r)^2 = 265/630$, giving $\sqrt{265} = (q + r)\sqrt{630}$. Therefore, (1) $\sqrt{271} + \sqrt{265} = (2r + p + q)\sqrt{630} = (1 + r)\sqrt{630}$ if random mating without selection; if not, the frequencies have been disrupted. The left-hand term equals 32.741 and the right-hand term is 32.649. Note that in the above expression, r in the left-hand term included heterozygous combinations and is computed from only the homozygous proportion or $\sqrt{57/630}$ in the right-hand term. The close fit suggests no meaningful shift from equilibrium. Consider the following sample:

Phenotypic Classes	Ab	aB	AB	ab	N
Frequencies	$p^2 + 2pr$	$q^2 + 2qr$	$2pq$	r^2	1.0
Sample Numbers	215	105	10	300	630

Proceeding as above, $\sqrt{300} = r\sqrt{630}$, $\sqrt{515} = (p + r)\sqrt{630}$, and $\sqrt{405} = (q + r)\sqrt{630}$. Therefore, (2) $\sqrt{405} + \sqrt{515} = (2r + q + p)\sqrt{630} = (1 + r)\sqrt{630}$. The left- and right-hand terms are 42.818 and 42.419, respectively.

These two values only differ by 0.399; however, the χ^2 test mentioned below indicates a larger difference need not occur in this expression for differences to be significant. If the left-hand term of expressions like (1) and (2) exceeds the right-hand term, then heterozygous combinations appear to be more abundant than random events predict. Fisher's material was largely composed of recessives, suggesting no selective disadvantage against them. The double dominants as a phenotypic class, AB or $2pq$, were very low in frequency. The following relationship exists:

$$(2r + p + q)\sqrt{N} - r\sqrt{N} = \sqrt{N}, \text{ or, in numbers,}$$
$$\sqrt{515} + \sqrt{405} - \sqrt{300} = 25.498.$$

Squaring the right-hand term gives 650.148 and the difference between 650.148 and 630 gives 20.148. This value approximated the missing number of class AB, assuming no reduction in fitness of the ab type relative to actual sample size. For the first sample, still with $N = 630$, $\sqrt{271} + \sqrt{265} - \sqrt{57} = 25.192$, and $(25.192)^2 = 634.637$. Approximately five individuals are missing from that sample for perfect agreement with expectation. In Chapter Two, gene frequencies were calculated for a sample having the same composition as sample one; sample two gene frequencies are obtained in a similar manner. Those frequencies allow a calculation of expected numbers. Note the following comparisons:

Phenotypic Classes	Ab	aB	AB	ab
Sample one				
Observed	214	208	151	57
Expected	212.6	206.6	152.9	56.9
Sample two				
Observed	215	105	10	300
Expected	201.6	89.5	23.9	308.7

A χ^2 test of sample two confirms that it is significantly different from expectation at the 0.001 level. A test of this type, or the one discussed below for linked genes, evaluates the likelihood of a neutral equilibrium. The deficiency of observed numbers in the AB class by gene frequency methods for sample two is 13.9, somewhat lower than Fisher's estimate of 20.1.

A potential error exists from the fact that the after-selection sample gives no information about the before-selection situation. If a non-neutral equilibrium exists, then gene frequencies do not change with selection. Successive before-selection samples are similar for gene frequencies and differ from after-selection samples only in genotype frequencies. The before-selection genotype frequencies are then predictable from the after-selection gene frequencies. This assumption, a reasonable one in view of the many cases of balanced polymorphism, was required by Fisher, and he continued his analysis to measure the approximate magnitude of selection. His expression to obtain expected numbers of class AB or $2pq$ was:

$$(\sqrt{p^2 + 2pr + r^2} + \sqrt{q^2 + 2qr + r^2} - \sqrt{r^2})^2 - (1.0 - 2pq)$$

which simplifies to $(p + q + r)^2 - (1.0 - 2pq)$, with all terms canceling but $2pq$. The numbers of sample one treated in this fashion are:

$$(\sqrt{271} + \sqrt{265} - \sqrt{57})^2 - (630 - 151) = 155.64.$$

The observed numbers and gene frequency estimates were 151 and 152.9, respectively, all rather close. The $2pq$ estimates for sample two are 30.04, 10, and 23.94, in the same order. Thus, only 33 percent (10/30.04) of expected numbers using Fisher's method exists in the observed sample. By such calculations, he found observed numbers in the actual samples were only 60 percent of expected values, and concluded 40 percent of the $2pq$ class of locusts were lost by selection. This treatment assumes no double dominant gametes, so the estimate is conservative. Since the data actually involve several linked genes, the minimal unit of selection is probably not a single locus. The period of selection measured by the single locus model is only one generation, wherein an s of about 0.4, for Fisher's data, exists for one suboptimal class. Such values are found in balanced polymorphisms, as discussed below. Fisher's paper provides a historical background on character frequencies largely unavailable in other groups. A current study of these populations would provide evidence related to the validity of his equilibrium assumption.

Two or More Autosomal Loci

The following methods compare genotype proportions to Hardy-Weinberg expectations by either gene frequency or Fisher's phenotypic proportion technique. Fisher's *Paratettix* material represented several linked loci, in which he recognized evidence of selection by what is now called linkage disequilibria. Fisher arranged the data for one sample (Tynan, Texas) as shown in the following contingency table:

| | Dominant Factor | | |
	D	Recessives	
Other Dominants	9	173	182
Recessives	149	888	1037
	158	1061	1219

The number of double dominants, 9, is less than one-nineteenth of non-D single dominants, 173, and the single dominant D combinations, 149, are just over one-sixth of the full recessives, 888. The $\chi_1^2 > 11.3$, confirming a significant difference in the distribution. The essence of the observation is that the dominant D factor is proportionately deficient with other dominants. The importance of this situation derives from the following property of two or more loci with random mating. Establishment of Hardy-Weinberg proportions in one generation applies to a two-locus system only in the unlikely case that only heterozygotes ($AaBb \times AaBb$) exist having all gene frequencies equal ($p = q = r = s = 0.5$). Otherwise, the approach to equilibrium requires a series of generations. At equilibrium, gamete frequencies satisfy the proportion $AB/aB = Ab/ab$, where $(AB)(ab) = (aB)(Ab)$. If the left and right terms are not equal, no equilibrium exists, and the difference, D, the coefficient of disequilibrium, equals $(AB)(ab) - (aB)(Ab)$. Fisher's data arranged in this form, and using proportions gives a D of -0.01197. Note that D is positive if the coupling phases are in excess, or negative if they are deficient. If frequencies of the genes A, a, B, and b are p, q, r, and s, then gamete frequencies at equilibrium equal the product of the appropriate gene frequencies. For instance, the frequency of AB at equilibrium equals pr. If AB_0, AB_1, and AB_e represent the frequencies of the AB gamete at the initial observation, one generation later, and at final equilibrium, respectively, the following expression gives AB_1:

$$AB_1 = AB_0 + C(AB_e - AB_0).$$

Similar expressions will give the AB frequency for each generation approaching the equilibrium. Thus, for obtaining AB_2, AB_1 is substituted for AB_0 in the above expression. The approach to equilibrium is thus determined by the constant C. Since maximal recombination occurs for independently segregating genes, and equals 50 percent, then C equals 0.5 for unlinked genes. If linkage exists, C is a smaller value equaling the crossover percentage. Consequently, gamete types will exist in proportions determined by the frequency of the genes for both linked and nonlinked loci. The only difference is the time required for the genes obtaining gametic equilibrium. The difference from equilibrium, D, is reduced by 0.5 each generation with nonlinked genes, and by 0.1 for two loci with 10 percent crossover.

Natural populations have had far more than adequate time to establish equilibria for even closely linked genes. In the absence of selection, the expectation consists of D equaling zero. The contingency table given above is probably the least complex method for questioning a statistical significance of the difference. When we find disequilibria, the most logical explanation consists of selection favoring a coupling or repulsion arrangement of genes. If the arrangement AB is favored by selection, the proportion of non-AB types produced by independently segregating alleles is higher than if AB are linked. Where selection favors a specific combination of genes, Fisher (1930a, 1958a) argues that an evolution of linkage will follow. The general condition of specific gene arrangements having higher fitness values than alternate conditions is now called co-adaptation, occasionally gametic phase inbalance, or using Mather's (1943, 1973) term, relational balance. These terms include both linked and nonlinked co-adapted genes, but most cases probably involve linkage disequilibria, since fewer ill-adapted gametes are produced each generation. Unless a population has recently received large migrational input from genetically different sources, linkage disequilibria probably represents selection. Theoretically, disequilibria can be established in one generation, although several generations are required to remove it. When it is found in a single sample, we have no insight on the number of generations of selection involved. This source of evidence has received limited application to date. Arnold (1968) used such data in interpreting samples of the snail *Cepaea nemoralis* from the Pyrenees. The background color of the shells is either yellow or pink and the shell's lip occurs as dark or white. These patterns are known to reflect different gene pairs and the gametic combinations were in disequilibria.

The contingency table analysis utilized above reveals only a significant absence of certain gene combinations, and, for the reasons given above, probably identifies linkages for most cases. If the entries are all sibs, this likelihood is much increased. If significant differences are not found by the test, the individual genotypes may still experience selection independently. Consider one gene pair before selection existing as $p^2(AA)$: $2pq(Aa)$: $q^2(aa)$, and a second pair as $u^2(BB)$: $2uv(Bb)$: $v^2(bb)$. Now let selection occur, with fitnesses for AA, Aa, and aa as $x : y : z$, and for BB, Bb, and bb as $x_1 : y_1 : z_1$, respectively. The frequencies of genotypes in the next generation's before-selection stage are given by expansion of the following expression:

$$[xp^2(AA) + y2pq(Aa) + zq^2(aa)] \; [x_1u^2(BB) + y_1 2uv(Bb) + z_1 v^2(bb)].$$

The ratio of A to a in class B equals its value in class b, the opposite of disequilibria; yet selection is present, but acting independently on the two gene pairs. The situation posed by fitness acting independently or otherwise on genotypes determined by two or more loci is discussed again in regard to the unit of selection.

Single before-selection samples scored for one gene pair give no evidence of selection since one random mating interval re-establishes the equilibrium. The one mating interval preceding a before-selection sample cannot remove effects of gametic disequilibria for two gene pairs, unless the phases were so close to equilibrium as to have no statistical difference. The evidence associated with disequilibria may therefore be sought in the analysis of before-selection samples.

Sex-Linked Loci

A single sample divides into two components, males and females, if sex-linked genes are concerned. In this case, the units compared are gene frequencies, p, q, etc., and the expectation, with no selection, consists of equal values of p for both sexes and for the total population. The rationale for this expectation appears in all books on population genetics and is briefly as follows: Expressions for p in males, females and for the total sample are:

$$p^m = \frac{N^A}{N^m} \; ; p^f = \frac{2(N^{AA}) + N^{Aa}}{2N^f} \; ; p^p = \frac{2(N^{AA}) + N^{Aa} + N^A}{2N^f + N^m}$$

where N^A, N^{AA} and N^{Aa} are sample numbers of type A males, homozygous dominant females, and heterozygous females, respectively, taking the male as the heterogametic, XY, sex. At equilibrium, the three p values are equal, but the nature of sex-linked inheritance prohibits establishment of equilibrium in one generation of random mating. Recall that the heterogametic sex, typically an XY male, passes his X chromosome only to his daughters, and receives his X from his mother. Consequently, male and female p values exhibit an oscillation from generation to generation on approaching the equilibrium point. The number of generations required for reaching equilibrium is directly proportional to the magnitude of difference between the p^m and p^f values. The population value of p doesn't change, only the male and female values. For instance, consider a population where males occur as 0.20 A and 0.80 a and where females occur as 0.20 AA, 0.60 Aa and as 0.20 aa. The p^m and p^f values are 0.2 and 0.5, respectively, while p^p is 0.4. In the next generation, the p^m is the preceding p^f value, 0.5, and the new p^f is a mean of the former values, namely 0.35. By six generations, most of the difference has vanished, with all values approximating 0.4. Of course, at equilibrium the genotypic proportions differ between the sexes. With $p = 0.4$, one-fourth of all males have the character and only one-sixteenth of the females, p^2, exhibit it. Thus, if such a system comes under selection, males are presenting a higher frequency of targets. As with linkage disequilibria, a sexual disequilibria in p values is most logically explained as a response to selection.

Metcalfe and Turner (1971) report an example of this type of selection evidence using gene frequencies of domestic cats in York, England. A sex-linked locus in cats exists for alleles determining black and orange coat colors. A large sample of cat data was assembled as the cats arrived by various means at the animal shelter over a 12-month period. Several loci were scored; however, the pertinent information here relates to the black and orange sex-linked alleles. The number of genes per sex and color (taken from Table 4, Metcalfe and Turner) appears in the following contingency table:

	Black Genes	Orange Genes	
Males	258	76	334
Females	583	121	704
	841	197	1083

The $\chi_1^2 > 4.2$, having a $P < 0.05$, indicating a significant difference in the distribution of black and orange alleles between the sexes. Actually, the authors use a form of the cross-product ratio in their analysis. They arrange the values in the following form:

$$\frac{x \text{ (No. Orange Genes in Males)}}{\text{(No. Black Genes in Males)}} = \frac{\text{(Total No. Orange Genes)}}{\text{(Total No. Black Genes)}},$$

where $x(76)/(258) = (197)/(841)$ and $x = 0.795$. This value shows the orange genes in males are in excess. The observed number of orange genes would have to be reduced by a factor of 0.795 to adjust the ratio so that x equals one. A test for the statistical significance of the x factor would follow Edwards (1965), mentioned above in connection with cross-product methods. The authors conclude that composition of the samples reflects the complex life experienced by urban cats. In some way, this life style involves selection, giving sexual disequilibria of the black-orange alleles. A number of other samples from different localities supported their conclusions.

The cats included adults and kittens, although probably most were old enough to classify the sample as an after-selection group. As noted above, differences in p values from equilibrium are not obscured by one mating interval unless the population is almost at the equilibrium point. Consequently, before-selection samples may be analyzed for selection evidence when using sex-linked genes, but such evidence expresses an unknown number of generations of selection.

Mother-Child Combinations

A special type of sample consists of mothers and their associated offspring all scored for their genotypes. The offspring are the individuals

Table 1. Incomplete family data with frequencies in idealized distribution

Parental genotypes and their frequencies		Frequencies of genotypes in offspring			
Female	Male	AA	Aa	aa	F_1 Totals
AA p^2	AA p^2	p^4			
AA p^2	Aa $2pq$	p^3q	p^3q		
AA p^2	aa q^2		p^2q^2		
AA p^2	1	p^3	p^2q		p^2
		$O_{A\cdot A}$	$O_{A\cdot Aa}$		$O_{A\cdot}$
Aa $2pq$	AA p^2	p^3q	p^3q		
Aa $2pq$	Aa $2pq$	p^2q^2	$2p^2q^2$	p^2q^2	
Aa $2pq$	aa q^2		pq^3	pq^3	
Aa $2pq$	1	p^2q	pq	pq^2	$2pq$
		$O_{Aa\cdot A}$	$O_{Aa\cdot Aa}$	$O_{Aa\cdot a}$	$O_{Aa\cdot}$
aa q^2	AA p^2		p^2q^2		
aa q^2	Aa $2pq$		pq^3	pq^3	
aa q^2	aa q^2			q^4	q^2
aa q^2	1		pq^2	q^3	
			$O_{a\cdot Aa}$	$O_{a\cdot a}$	$O_{a\cdot}$
Parental totals		p^2	$2pq$	q^2	1.0
		$O_{a\cdot A}$	$O_{a\cdot aa}$	$O_{a\cdot Aa}$	

[1] Assumed to be random.

for whom selection affecting the parents is expressed. The relative proportions of different offspring genotypes from each maternal genotype may be predicted, assuming that she has mated randomly. The analysis having most promise for natural populations involves incomplete family data, IFD, as suggested by Cooper (1968). Presumably, the method is applicable to any species in which the young and mother may be associated and multiple matings are absent. Examples include field-collected gravid or pregnant females that will subsequently produce young in the laboratory, birds tending nestlings, marsupials with young in the pouch, etc. The random mating assumption brings genotypes together for mating in numbers proportionate to their frequencies. The product of these frequencies will be the proportion of their offspring in the population's total F_1 generation. This fact can be seen from Table 2 of Chapter Two and again in Table 1 of this chapter.

The expected proportion of F_1 with a given genotype from a given parental genotype may be obtained by summing the possible combinations. For instance, AA F_1 from AA mothers should be $p^4 + p^2 q$, or p^3 ($p + q$), or p^3 of the total p^2 offspring of such mothers. The letter O signifies offspring, the subscripts give genotype of parent-offspring, respectively. Namely, $O_{A \cdot Aa}$ identifies Aa F_1 from AA parents. Expected F_1 genotype proportions are thus available for each parental genotype and the marginal totals for F_1's and parents equal $p^2 + 2pq + q^2$ as expected. Cooper observes that algebraically, the following expression is a 1:1 ratio:

$$\frac{O_{Aa \cdot A} + O_{Aa \cdot a}}{O_{Aa \cdot Aa}} = 1.0.$$

If a 1:1 relationship does not exist, the terms are not in the predicted frequencies. This condition is assumed to come not from nonrandom mating but from the fact that parental genotypes were not in predicted equilibrium proportions. This observation applies to an after-selection group (the parents) and constitutes evidence of selection. The ratio is tested by a χ^2 in which expectation places the two terms in the expression as equal. Cooper notes that additional tests are possible from the data, for instance, the ratios:

$$\frac{O_{A \cdot A}}{O_{A \cdot Aa}} = \frac{O_{Aa \cdot A}}{O_{Aa \cdot a}} = \frac{O_{a \cdot Aa}}{O_{a \cdot a}} = \frac{p}{q}$$

These ratios are termed *gametic gene ratios*. One possible complication leading to significant departures from expectation is the influence that subdivision of a population can have on the genotype frequencies, the so-called Wahlung Effect, a simulation of inbreeding by increasing frequencies of homozygotes. The appropriate χ^2 test takes all three ratios as equal for the expectation with two degrees of freedom.

An example of this procedure exists in Blue Grouse (*Dendragapus obscurus*) data in Vancouver Island, British Columbia (Redfield, 1973a). The *Ng* locus was scored for the *M* and *S* alleles using blood samples and electrophoresis. The sample consisted of chicks, usually less than ten days old, presumably representing a before-selection sample. For the years 1969, 1970, and 1971, the expected 1:1 ratio was 108:65. The χ^2 for this distribution is 10.2 with $P < 0.001$ showing a clear case for selection. The departure from 1:1 involved more homozygotes than expected, and Redfield predicts that fitness changes at a later stage of the life cycle to favor heterozygotes. This conclusion is necessary, since neither the *M* or *S* alleles appear to be moving toward fixation. If offspring are scored after experiencing a reasonable period of natural habitat life, a departure from 1:1 could reflect effect of selection from two generations.

Another approach uses only the offspring of recessive mothers, and has been applied to the snail *Cepaea nemoralis* by Murray (1964). In this snail, yellow shell color is recessive to pink, and a sample of recessive types fertilized in the field were collected and later produced 35 broods of young in the laboratory. If random mating was occurring, the expected F_1 from this recessive sample can be depicted as follows:

		Offspring		
Mating Types	Frequency	AA	Aa	aa
$aa \times aa$	$q^2 \times q^2 = q^4$			q^4
$aa \times Aa$	$q^2 \times 2pq = 2pq^3$		pq^3	pq^3
$aa \times AA$	$q^2 \times p^2 = q^2p^2$		p^2q^2	
			pq^2	q^3

The proportion of dominant (pink) in the offspring sample should be $pq^2/(pq^2 + q^3)$, or p, the gene frequency of the dominant allele in the population providing the sample. In the population sampled, Murray found 0.187 and 0.813 percentages of pink and yellow phenotypes, respectively; therefore, with random mating and no selection, q equals 1.0 $- \sqrt{0.813}$, or approximately 0.10. Of the 1,419 progeny scored, 262 were pink and 1,157 were yellow. Judging from the 10 percent expectation for q, 141.9 and 1,277.3 would be pink and yellow, respectively. A χ^2 comparison of the observed and expected numbers reveals a $P < 0.01$ for explaining the deviation by chance. The interpretation that yellow snails fertilized by pink mates leave larger numbers of pink progeny than expected is complicated by two factors. The expectation rests on parental genotypes having been in Hardy-Weinberg proportions. If inequitable reproductive success occurs in one generation, it probably occurs, more or less, each generation. As a consequence, the genotype frequencies used above for parents may not be correctly represented. Thus, the frequency of mating types shown above would change, probably involving a shift toward higher proportions of pink phenotypes and, as a result, more pink progeny. The principle objective of Murray's study was to demonstrate multiple matings between the snails and the possible effect of sperm storage in counteracting effects of a low effective population size that might otherwise be envisioned. In the above table, even if previous selection was absent, the matings represent combinations between only two parents and do not recognize multiple matings between genetically different mates. The high frequency of recessive phenotypes suggests that most second and third matings would have been with aa or yellow individuals. This effect would increase the frequency of yellow progeny beyond the expectation shown above. The expected proportion of recessive F_1 based on the single-mate pattern without previous selection is $q^3/(pq^2 + q^3)$, or 0.9. Murray's analysis predicted about 2 to 3 matings per individual, and with the high observed frequency of recessive phenotypes, the proportion of yellow progeny would be expected to exceed 0.9. The observed proportion is 1,157/1,419, or 0.815, and it is unlikely that nonequilibrium genotype frequencies of the parents could account for this much difference. Thus, one feature, reproductive inequalities of parents, predicts an excess of pink progeny, and the second feature, multiple matings in connection with observed phenotypic proportions, predicts an excess of yellow progeny. The data are not compatible with expectation involving only random but multiple matings; thus, selective advantage for the pink snails, probably for reproductive differences, is a reasonable likelihood. The high observed frequency of yellow shells, on the other hand, suggests balancing selection, perhaps in terms of survival, favoring yellow shells. Estimating reproductive inequalities by associating success with a phenotype score only may not reveal the complexities involved. Heterozygous pink shells in *Cepaea* may, for example, have higher correlation with maximal reproduction than the full phenotypic class of pink shells. In fact, this condition was suggested by Murray's data, since no broods occurred with only pink shells, presumably the F_1 expected of homozygous dominant pink mates.

A related approach for testing equilibria exists with Snyder ratios, measuring proportions of recessive offspring from dominant X dominant and and dominant X recessive matings (Li, 1955). In this case, phenotypes of both parents must be known, limiting its use with most nonhuman populations. On the other hand, the ratios suggested by Cooper require an ability to distinguish the heterozygote. Recognition of heterozygotes is frequently impossible, since most loci seem to have dominant alleles, and especially when not scoring by molecular properties (for example, electrophoresis). If the IFD entries above are pooled into phenotypically indistinguishable values but a separation based on the mother's phenotype is retained, four terms remain. These terms are placed in the following table:

Offspring

Mother	$A-$	aa	Totals
$A-$	$p^3 + 2p^2q + pq$	pq^2	$p^2 + 2pq$
aa	pq^2	q^3	q^2
Totals	$p^2 + 2pq$	q^2	1.0

This arrangement gives another form of mother-child combinations, and the expected ratio of recessives to total F_1 from dominant mothers is $pq^2/(p^2 + 2pq)$, or $q^2/(1 + q)$, and in the F_1 from recessive mothers, the value is q^3/q^2, or q. To test for equilibria, an estimate of q is obtained from the data's q^2. A sample of human data from Snyder (1932) adjusted for mother-child analysis is given below:

Offspring

Mother	$T-$	tt	Totals
$T-$	1,170	269	1,439
tt	241	357	589
Totals	1,411	626	2,028

To obtain q^2, the two marginal totals shown as proportionate to q^2 are added to use all information and divided by twice the sample size ($1,215/4,056 = 0.2995 = q^2$) and $q = 0.5473$ and $p = 0.4527$. Note that the estimate combines recessive offspring segregating from both dominant and recessive mothers. The expected proportion of recessives from dominant mothers is 0.1936, and the observed is 0.1869. The expected and observed values for recessive mothers are 0.5472 and 0.6061, respectively. Using the calculated q^2 to obtain expected numbers, the following comparison evaluates the differences:

	$T-$	tt
Observed F_1	1,411.0	626.0
Expected F_1	1,426.9	610.1

The $\chi_1^2 > 0.43$, with a $P > 0.3$, showing no significant departure from the expected equilibrium. As with IFD, the proportions assigned to each phenotypic class of F_1 are derived by assuming the mother's genotypes to be in equilibrium with her gene frequencies.

If the data are not in equilibrium and F_1's are truly a before-selection group, selection experienced by the mothers presumably is the distorting agent of genotype proportions. However, a heavy selection would have to be operating to change these values sufficiently in one generation for statistical recognition. This restriction applies also to IFD ratios. The so-called neutral equilibria found by mother-child comparisons may only reflect small selection coefficients, as mentioned elsewhere. Sperm storage should not invalidate the method unless differential longevity of sperm occurs, a situation equivalent to gametic selection.

Within-Generation Multiple Samples

The selection period reflected by such data is only one generation, and as suggested earlier, our tests for such periods are insensitive to small or even moderate selection coefficients. The data are analyzed by one of two methods. One approach involves comparing survival of different classes as they exist in the natural environment. A second method consists of transferring an age-cohort to the laboratory, where its subsequent development occurs in an environment shielded from, or with different, selection. The character frequencies existing at a later age in the laboratory stock are then compared to frequencies occurring in the comparable age of individuals from the natural environment. Frequency differences found by either method are interpreted as evidence of differential survival. The following studies give examples where recognizable differences in survival occurred for major gene and polygenic expressions.

Kettlewell (1973) summarized earlier work on industrial melanism in England for the moth *Biston betularia*. The species varies for genes at a single locus, producing a dark form, *carbonaria*, and a typical morph. The moths perch for lengthy periods on tree trunks, and the typical morph closely resembles the lichen-covered bark of trees. In and near cities, the burning of coal associated with heavy industry has darkened many surfaces, including the tree trunks. In such areas, the melanic, *carbonaria*, occurs in high frequencies. Birds use the moths as food and utilize vision in locating their prey; therefore, the hypothesis envisioned higher survival of melanics in industrialized habitats. About 2,000 moths were marked and released in areas where pollution had darkened trees and in more natural areas having lichen-covered bark. The proportions of the two morphs released constituted expected frequencies among recaptures. The morphs exist in both sexes, but females rarely fly or do not assemble in the collecting traps, so the data represent conditions for males. Only 50 percent of the expected typicals were recaptured in polluted areas and 67 percent of expected *carbonaria* were recovered in the natural woodland. In each case, the more conspicuous morph was recovered in significantly lower proportions. Kettlewell also has obtained evidence of bird predation by direct observation. The study stands as a classic demonstration of natural selection where the selective agent is known. In such work, one assumes that nothing other than availability influences chances of recapture, and if equating results to actual fitness, as Haldane (1956) did, reproductive abilities for the two morphs are assumed equal. These assumptions are probably valid for *B. betularia*. Other studies have refined analysis of recapture data and are discussed below relative to fitness estimates.

Bell (1973) followed character frequencies in a natural environment by measuring quantitative traits at different periods in an age-cohort of smooth newt larvae, *Triturus vulgaris*. He found that variation reduced with advancing age. The data were analyzed by several statistical techniques, and he concluded that developmental canalization, mutual growth inhibition, and statistical artifact were not significantly involved. Thus, by negative argument, he contributed the reduced variation to selection. A more direct association of quantitative traits with a selection agent occurs in the study by Bumpus (1896) on English Sparrows, an early classic on natural selection. He compared measurements of nine morphological traits for specimens dying in a severe winter storm against those of survivors, concluding that differences existed and selective elimination was responsible. Other authors have reanalyzed the Bumpus data (Harris, 1911; Johnston, et al., 1972; O'Donald, 1973), also confirming a strong case for selection. A period of selection operating on a recognizable age group may be expected to reduce its variability. This effect is most easily recognized by an F test on the before- and after-selection variances. Further analysis involves the selection intensity and variance of fitness discussed below.

The second method is exemplified by Camin and Ehrlich's (1958) paper on water snakes, *Natrix sipedon*. The snake population occurred on islands in Lake Erie where individuals live among flat limestone rocks forming the shore line. The snakes are frequently exposed, and the presence of several vision-oriented predators (gulls, hawks, etc.) suggests a likely case of observable selection. The island population tends to be pale and less banded in pattern than mainland counterparts, presumably a cryptic property derived on the rocks. The authors collected pregnant females and subsequently obtained young snakes equivalent to a before-selection sample. The banding pattern was much more variable in this sample than in adults. The difference involved larger numbers of completely banded, dark patterns in the young snakes. Actually, the sample of adults compared to the sample of young snakes born in captivity was not the same age-cohort; however, the adults were apparently stable in pattern frequencies between successive generations, validating the comparison. Some migration from the mainland to the islands exists, and the authors suggest this input of dark-banded attributes may prevent development of even more differentiated island populations.

A similar procedure was followed by Dowdeswell (1961) with the butterfly *Maniola jurtina*. The species possesses from 0 to 5 spots on the underside of the hindwings, and has been the object of several studies in ecological genetics. Dowdeswell collected larvae, apparently varying in age, and reared them into adults in the laboratory. When comparably aged butterflies developing in the natural habitat appeared, samples were taken for comparison. No significant differences appeared in spot frequencies between male samples or late-season female samples; however, May and early June female samples did vary significantly. Thus, he obtained evidence of a seasonal difference in relative survival of female types and also a sexual difference in fitness. The selective agent was thought to be, in part, a hymenopterous parasite.

Redfield (1973b) presents a modification of analyzing data that appears to reflect within-generation samples. He compared contemporaneous samples from different habitats using the *Ng* locus in Blue Grouse, *Dendragapus obscurus*. The grouse habitat undergoes successional change associated with forestry practices. The different stages exist simultaneously and can be recognized with reasonable objectivity. Redfield's data consist of "yearlings." The range in age of such grouse is not clear, but such samples represent individuals that have experienced a not too variable period of life, and thus provide similar after-selection comparisons. The expected situation, assuming the whole population mates randomly, consists of equal frequencies in all successional stages. The factor *NgM* was significantly higher in frequency for samples taken from one-year-old habitats. The disproportionate frequency was interpreted as evidence of selection. The habitat stage with the higher *NgM* frequency has a one year duration and selection identified by this method is presumably a response during the year. The allele may achieve balance in part by an oscillation with habitat succession. Redfield notes that the grouse are more numerous in one-year-old habitats than other stages and associates the selection agent with density, though other interpretations exist. If the successional stages are creating subdivision relative to mating, an equal expectation over all habitats is no longer valid.

The morphs of *Biston betularia* and the Ng factor in the grouse are clearly segregating genetic units; however, traits scored by Bell, Bumpus, Camin, Ehrlich, and Dowdeswell were phenotypic variations with unknown genetic components at the time of the studies. Various *Natrix* banding patterns apparently appear in progeny of single litters, suggesting segregation of genetic factors. Characters can change expression during ontogeny, complicating the comparisons. For instance, Howard (1940) describes the variable change in color patterns of the isopod *Armadillidium vulgare*. The change varies with age and sex in that species, and this possibility in reptiles has received less attention. Since Dowdeswell's work, McWhirter (1969) analyzed heritability of spotting in *Maniola*, with the interesting finding of approximately 0.14 and 0.63 in males and females, respectively. The higher heritability in the sex having greater response to selection is as expected, since higher heritability values reflect greater genetic variability that may be sensitive to selection. The heritability concept is discussed more fully in a later chapter. The quantitative traits measured by Bell and Bumpus have not been examined genetically, although they doubtless involve genetic components. The studies on sparrows, snakes, butterflies, and newts appear to reveal stabilizing selection. Kettlewell's data on the peppered moth probably apply to a genetic polymorphism near balance and Redfield's study seems reasonably explained as an ecologically balanced polymorphism. The latter two studies also revealed rather high s values, and such estimates are commonly found in balanced polymorphisms.

Between-Generation Multiple Samples

Frequencies may be recorded by either gene, genotype, or phenotype proportions. Note, however, that two samples scored by phenotypic proportions where dominance occurs may involve changes that are not revealed. The proportions of $(D + H) + R$ may exist as $(0.50 + 0.25) + 0.25$, $(0.05 + 0.70) + 0.25$, etc., where both appear as $0.75 + 0.25$ but have q values of 0.625 and 0.850, respectively. For this reason and others, a knowledge of the genetic basis for the scored variability is required for a meaningful analysis of between-generation data. These analyses assume undisturbed, random mating and both before- and after-selection comparisons, or equivalent situations will reveal changing frequencies. Successive before-selection samples reveal selection changes occurring in the generation of the first sample, while successive after-selection samples measure selection during the generation of the second sample. Also, a before-selection sample in generation 0 compared with an after-selection sample in generation 1 is equivalent to before-selection comparisons between generations 0 and 2. The p value of the after-selection sample equals the p value in generation 2's before-selection age group, assuming no gametic selection. Of course, two generations contribute to the observed selection magnitude. Prout's studies mentioned earlier indicate the practical difficulties in applying these properties to actual data. For our purposes, we must realize that the following studies fall short of estimating the net fitness by an unknown error.

Between-generation data are required to document directional selection. Such selection is ideally observed by following gene frequencies through consecutive generations. Small s values, not recognizable in methods measuring effects of one generation, are more likely to be documented by this approach but may nonetheless be difficult, as shown. Another approach consists of comparing populations separated by a large number of generations. This method is actually a small-scale version of the way paleontologists consider data, since the real number of generations is usually unknown. Two examples of treating consecutive generation data are given, followed by a case of widely separated generations.

Wright and Dobzhansky (1946) reported efforts to duplicate in the laboratory changes observed in nature in *Drosophila pseudoobscura*. In the paper, they present a test for selection applicable to comparing either before- or after-selection samples. The data actually analyzed came from population-cage experiments in which the introduced flies carried chromosome inversion frequencies of the authors' choosing. The inversions consist of different gene arrangements on the third chromosome pair, and are scored in stained salivary gland preparations of larval tissue. The cultures were maintained at different temperature regimes. Each subsequent sample consisted of egg batches taken over six consecutive days, each giving typically 25 larvae, for a total 300 chromosomes. These samples appear to reflect before-selection conditions for the populations. The inversions are discontinuous traits and are transmitted to subsequent generations in the same fashion as alleles. Consequently, the Hardy-Weinberg model can be used to express their frequency as it was by Fisher in his adoption of it for analysis of the *Paratettix* data.

The rationale for their method observes that as selection changes a gene's frequency, the relationship of Δq to q is curvilinear (Figure 2,

Chapter Two). In directional selection, Δq is small while q is small, and becomes maximal at intermediate q values, followed by low values again as q approaches one. In such selection, Δq equals zero only when q equals zero or one. When heterozygotes are favored, Δq will equal zero at q values of 0, 1.0, and an intermediate point. The latter case is of course the most interesting value. Depending on the relative fitnesses of p^2 and q^2, the Δq curve becomes S-shaped and crosses the horizontal axis representing q at the equilibrium point. As q rises or falls from this point, the values of Δq change and push q back to its equilibrium point. Consequently, if q expresses a significant response to selection, q and Δq are both changing in a functionally related fashion and the regression of Δq on q will not give a regression coefficient of zero. Likewise, a Δq differing from zero testifies to selection. A $\overline{\Delta q}$ of zero existing for an intermediate q value may, however, also approximate zero due to + and − values to either side of the equilibrium point. In this case, the regression coefficient will still differ from zero.

Nonetheless, the method requires appreciable data before statistical evidence is generated. As an example, one analysis by Wright and Dobzhansky (1946), in which the full complement of data revealed selection, involved 30 generations at 25°C. The generations consist of pooling results of seven subtests. The longest series of successive generations was eight, and occurred in the subtest identified as "experiment 18." Expressing the Chiricahua frequency as q, the data for that experiment appear in the following table:

Generation	q	Δq	Summations
1	0.365		$\Sigma q = 2.242$, $\Sigma q^2 = 0.6771$
		+0.028	
2	0.393		
		−0.056	
3	0.337		
		−0.030	
4	0.307		$\Sigma \Delta q = -0.1850$, $\Sigma \Delta q^2 = 0.0182$
		−0.050	
5	0.257		
		−0.067	
6	0.190		$\Sigma q \Delta q^2 = -0.0731$
		+0.040	
7	0.230		
		−0.067	
8	0.163		
		+0.017	
	0.180		

The initial q was 0.365 and proceeded in eight generations to a value of 0.180, with the associated Δq values shown. The regression coefficient of Δq on q can be taken from:

$$\frac{N \Sigma q \Delta q - (\Sigma q)(\Sigma \Delta q)}{N \Sigma q^2 - (\Sigma q)^2}$$

where N equals generation number. This value solves to -0.435. The 95 percent confidence limits for the coefficient requires calculating standard errors of the estimated Δq and of q by conventional methods, and using Student's t distribution with $N-2$ degrees of freedom. One then obtains -0.435 ± 0.654, clearly not significantly departing from zero. The Δq is -0.0231, and its 95 percent confidence limits give -0.0231 ± 0.032, again showing no significant average rate of change from zero. On the other hand, the rather progressive change in q from 0.365 to 0.180 intuitively suggests some process like selection at work. Examination of consecutive samples is also instructive. Two such samples representing only one generation of possible selection may be compared for a number of gene arrangements. The frequency of chromosomes for the first two generations after initiation of experiment 18 were as follows:

Gene Arrangements

	ST	AR	CH	N
Generation 1	0.333	0.273	0.393	
	(99.9)	(81.9)	(117.9)	300
Generation 2	0.377	0.287	0.337	
	(113.1)	(86.1)	(101.1)	300

The approximate numbers involved are the values shown in parentheses (actual numbers were not published). These numbers compared by a χ^2 test using a 2 X 3 contingency table are not significantly different, and similar results are found with other pairs of consecutive generations. These two generations represent part of the period when *CH* went from 0.393 to less than 0.20, presumably by directional selection toward a balanced polymorphism value. This observation underscores the fact that samples representing only one generation of selection may appear to be in equilibrium, since the Δq per generation is too small for statistical recognition.

Fisher and Ford (1947) presented data on eight successive generations for the moth *Panaxia dominula* in England. The samples consisted of adults having a mean life expectancy of just less than seven days. The values are thus potentially after-selection in time, and selection may have operated against either larvae or adults, or both. A single locus, affecting the expression of spots and extent of black pigment on the wings, was followed through this period. The common homozygote, *dominula*, the heterozygote, *medionigra*, and the rare homozygote, *bimacula*, may all be recognized phenotypically. The gene determining the rare homozygote fell from approximately 10 to 5 percent over the eight year observation. The change was too great to explain by typical mutation rates, and the colony was judged to be isolated relative to possible migration. The authors were thus distinguishing between an explanation based on random genetic drift and one based on selection. The question is, then, could the sample sizes obtained year by year in relation to the total population available each year involve random deviations of gene frequency large enough to explain the observed changes? Their capture-recapture data revealed numerical fluctuations, but the lowest density observed approximated 1,000. They used the value of 1,000 as the total population each of the eight years, allowing a higher likelihood for drift in their calculations than actually occurred. The standard deviation for a proportion as low as the observed gene frequency is not normally distributed; therefore, the authors converted frequencies by an angular transformation, $p = \sin^2 a$. These values and variances of sample size relative to available number were then analyzed by matrix algebra methods. The concluding result produced a $P < 0.01$ that the gene frequency changes were random.

Wright (1948) questioned Fisher and Ford because of the unknown population numbers prior to the study, possible fluctuations of selection coefficients, and possible nonrandom properties of oviposition and subsequent mortality. Ford (1975) summarized evidence that largely invalidated the objections, but this study and the previous one with *Drosophila* demonstrate the rigor required in documenting selection for between-generation data.

An example of comparing widely separated before-selection samples also involves chromosome inversions of *Drosophila pseudoobscura*. The data exist for a number of populations from scattered sites in the American southwest. Dobzhansky (1958) summarizes the data from the 1940's to 1957. A single larva (two third chromosomes) from the progeny of each field-collected female was scored. Field-collected males, when used, were mated to known laboratory stocks, and progeny of those crosses analyzed. Fully grown larvae are usually used, and one assumes that the laboratory culture was free of selection or, at least, differed significantly from selection in the field for the period between oviposition and scoring. The number of field-collected females was sufficiently high to make sampling error unlikely. These data reveal large changes in frequencies. For example, the so-called Standard Arrangement, ST, changed in California populations from 28.6 to 45.2, 26.9 to 57.0, 43.7 to 61.0, and 13.8 to 25.5 percent at two Yosemite Park sites, a population near the coast at Guerneville, and at Mount San Jacinto, respectively. Changes involving similar differences were found in other arrangements in sites east to Texas. Dobzhansky did not present statistical confirmation of significant differences; however, the magnitude of observed changes, having samples of well over 100 individuals, appear to speak for themselves. The samples were separated by a large number of generations for many sites, and no evidence exists on the nature of inversion frequencies for most intervening years or of the possible advantages of a specific gene arrangement. Dobzhansky considers several explanations, concluding that only selection could account for the data. Geographically close populations were found to change in similar ways, supporting his interpretation.

The *Panaxia* data and the *Drosophila* population-cage data both reflect short intervals of directional selection. The *Drosophila* were seen to establish a balanced polymorphism, and the *Panaxia* genes moved toward an oscillating state, although this was not clearly evident at the time of Fisher and Ford's publication. Directional selection has been identified as transitional polymorphism, both terms describing the progressive rise or fall of a gene's frequency over successive generations. Analyzing stabilizing selection by between-generation data would involve, in simplest form, comparing a before-selection sample with an after-selection sample of the following generation, thus pooling two periods of selection. Assuming no sampling artifacts arose by the method and that the before-selection conditions are similar for both generations, the data would reflect a mean of the two periods. If significant overlap of generations exists, such estimates may reflect conditions the population actually experiences.

STUDIES IN THE LABORATORY

Moving to the laboratory to study natural selection is a retreat from the uncontrolled variables and difficulties of sampling in nature. The basic assumption is that differential success of genotypes in the laboratory environment will also operate in the field, though the direction and magnitude may be unpredictable. This assumption is reasonable since the components of the laboratory environment also occur, in variable form, in nature. Negative findings in the search for selection in the laboratory are difficult to interpret. The laboratory is always shielded, to some extent, from known and unknown variables that exist in nature. What may be an important agent of selection in the field may be totally lacking in a laboratory colony, and such agents may be difficult to recognize. Thus, characters cannot be validly labeled as neutral from negative laboratory evidence.

The study by Wright and Dobzhansky discussed above is an example of transferring a natural population into the laboratory, where selection agents are then varied and population responses analyzed. More direct artificial selection consists of intentionally choosing specific parental phenotypes as parents for each successive generation, imposing an intense level of directional selection, and then noting the population's ability to respond. Such examples of artificial selection are discussed in detail by Falconer (1960). The essential observation for the present discussion consists of the fact that nearly all selection efforts meet with some success, showing the sensitivity of chosen characters to the selection process. The traits usually studied this way have quantitative expressions and are discussed further in Chapter Seven. Laboratory investigators have also pursued the quest for selection evidence in other ways.

The reproductive component of fitness is difficult to disentangle from an observed fitness difference. The following approach exemplifies an effort to specifically document the reproductive component. Anderson and Watanabe (1974) used the rationale with *Drosophila pseudoobscura* concerning the inversions of the third chromosome. As an example of the method, ST/ST and ST/AR females were mated with only ST/ST males. The gametic contributions were expressed as follows:

	ST/AR	ST/ST
Parental Female Type		
Parental Female Frequency	P	$1 - P$
Fitness	1	w
	P	$(1 - P)w$
Gametic Contributions	$\dfrac{P}{P + (1 - P)w}$	$\dfrac{(1 - P)w}{P + (1 - P)w}$

Since males contribute only ST chromosomes, the expected ST/AR frequency in the F_1 is half of the frequency of parental ST/AR females. Adjusting this value for a total of one gives $(P/2)/[P + (1 - P)w]$. The ST/ST frequency in the F_1 consists of the sum of frequencies for the remaining half of ST/AR and all of the ST/ST female parents, the value being $[P/2 + (1 - P)w]/[P + (1 - P)w]$. The observed F_1 frequency of ST/AR, H, equals $(P/2)/[P + (1 - P)w]$, giving a solution for w as $w = [P(1 - 2H)]/[2H(1 - P)]$. One such test used 500 females (50 ST/ST and 450 ST/AR) and 500 ST/ST males, so that P and $1 - P$ equal 0.9 and 0.1, respectively. Among 200 F_1 larvae, 130 were ST/ST and 70 were ST/AR, so that H equals 70/200, or 0.35, and w is 3.86. With no differential reproduction, w would be one for both karyotypes, and the expected ST/AR and ST/ST frequencies are 0.45 and 0.55, respectively. Thus, for F_1 ST/AR and ST/ST types, the observed and expected numbers are 70, 130 and 90, 110, respectively. A χ^2 value of 7.6 results, indicating a $P <$ .01 for chance deviations as large as observed. The logical conclusion is that differential reproduction exists. Anderson and Watanabe also found w to vary with frequencies of female parental types. Of course, the computed w cannot be extrapolated to natural populations, but the potential for such a fitness component is verified and it could be equal to, or more critical than, differential survival. Clearly, this component of fitness is not recognized by within-generation samples. After-selection samples, as defined for survival comparisons, become before-selection relative to reproductive inequalities. However, the earliest samples available for a generation usually involve, in practice, a possibility for differential survival.

RESTRUCTURING GENOTYPES

If a species' endowment has been molded by selection, then a restructuring of this endowment should generally reveal nonadaptive expressions. The following examples support this conclusion by noting response of the background genotype. These studies restructured genotypes by: 1) inbreeding, 2) artificial selection, and 3) combining genes from geographically distinct races. Genotypes automatically tend to restructure themselves by segregation, and recombination and genetic controls on this type of change also relate to selection evidence.

While sex-determining mechanisms are varied, the events share a point in common, namely, the ability to switch development into either a normal male or female. Development requires a balanced operation of the complete genotype, although the switch may be something like the regular distribution of only the X and Y chromosomes at meiosis and fertilization. If the background genotype can be varied experimentally, while the X-Y balance remains unchanged, the effects on development should be observable. Mather (1948) describes such a test on the plant *Lychnis dioeca*. Background genotypes were restructured by the inbreeding, leading to an imbalance during development resulting in hermaphroditic males. Presumably, selection in the natural population developed a switch mechanism, with its threshold sensitive to the average degree of heterozygosity maintained by the natural mating system. When this heterozygosity decreases under inbreeding, the switch becomes abortive.

Another likely example of background genotypes responding to selection concerns the evolution of dominance. Genotypes that are directly scorable in the phenotype, namely not polygenes, are often called major genes. Traits controlled by major genes are the expression of the dominant allele for a large percent of known cases. Mutations generally appear with expressions not well adjusted to the integrated gene pool, thus having low fitness expression, but the expression is generally hidden by the property of dominance in the more common "wildtype" allele. Presumably, the wildtype allele also came into the species population as a mutant at one time. The process of developing the dominance has been actively pursued, generally in terms of natural selection. Fisher (1928) suggested that a gene's expression is not independent of its neighboring genes and that such neighbors collectively modify a gene's net expression. Thus, the expression of a gene is largely determined by a set of modifier genes. A gene may thus be considered in terms of its independent expression, the major gene property, or in terms of a modifier for another gene's expression. If a locus A_1A_1 mutates to A_1A_2, which expresses in part the less fit mutant allele, the following generations (if A_2 is not lost) will consist of far more A_1A_2 combinations than A_2A_2 because of the very low frequency of A_2. If another locus exists, M_1M_2, with modifying effects on the A locus, and with variability (thus recognizing existence of both M_1 and M_2), theoretically, it follows that M_1 and M_2 will have different effects on the fitness of A_1A_2. If M_1 enhances A_1A_2 to phenotypically copy A_1A_1, while M_2 allows expression of A_2, then selection will develop an advantage for $A_1A_2M_1M_1$ types, or for $A_1A_2M_1-$ if M_1 has existing dominance over M_2. Since A_2's expression is assumed, initially at least, to have low fitness, then M_1 will increase in frequency and eventually all combinations of $A_1A_2M_1-$ will phenotypically appear as A_1A_1 where A_1 has developed dominance over A_2. The initial presence of both M_1 and M_2 is not unlikely in view of the many known polymorphisms. In this case, the selection coefficients affecting an initial balance in M_1 and M_2 would change and a new balance would result, or perhaps M_2 would be lost. Reflection on this explanation of dominance reveals a number of problems. These difficulties with the theory are given below, but the nature of supporting evidence is first outlined.

The experimental data seem to support Fisher and fall generally into two designs. First, selection over succeeding generations may be exercised on variable heterozygote expressions to observe what evidence, if any, may develop for heterozygous genotypes copying homozygous classes. Secondly, heterozygotes of within-race crosses may be compared with similar genotypes obtained from interracial crosses in which different sets of modifiers may well be expected to determine dominance of one major gene.

An example of the first procedure was reported by Ford (1940) for the moth *Abraxas grossulariata*. The usual wing color is white or pale cream but a less common form, *lutea*, exists, having a deep yellow color. The white and deep colors represent homozygotes for a single gene locus, and the heterozygotes are light to rather medium yellow. Ford crossed a normal white form to a heterozygote and, from the heterozygous F_1, he used light and medium-dark yellow heterozygotes as parents for a light and dark line of matings, respectively. This procedure continued for the offspring of each generation from both light and dark stocks. As soon as the third and fourth generations, fully white and fully deep yellow heterozygotes were obtained from the light and dark lines, respectively, as shown by test crosses. This case is a particularly rapid response by heterozygotes in which the ability to undergo dominance modification is seen for both alleles. *Drosophila* workers often observe the gradual loss of distinct expression for a mutant when the flies are cultured over successive generations as a single population, apparently a similar observation. Clarke and Sheppard's (1960) work with African races of the swallowtail butterfly *Papilio dardanus* illustrates the breakdown of dominance in interracial crosses, presumably due to different modifier systems in the different races. Thus, evidence exists that modifiers do influence the degree of dominance, and dominance may be significantly changed over a short number of generations.

A close look at the dominance modification theory reveals some rather deep problems that require additional assumptions. Haldane (1930, 1939) drew attention to the restriction that fitness requires. The fitness, w, of *AaMM* or *AaM−* must exceed the selection coefficient, s, operating against *aamm* or *aaM−*. If this difference is absent, then M is at a disadvantage when a is at even modest frequencies. Also, the fitness of *AaMM* or *AaM−* must equal or exceed the fitness for *AAM−* or directional selection could increase the frequency of M without a necessary influence of dominance. Both of these properties require a rather high frequency of the *Aa* genotype before they can become meaningful, and this requirement is the principle problem, i.e., how does a mutant with poor fitness become high enough in frequency for the proposed selection to function? Wright (1929a, b) notes that since *Aa* initially is very rare, the modifier, M, converting its expression to *AA* may also be rare and have been lost by drift before the time of its possible advantage. Crosby (1963) has more recently argued that examples of dominance modification, such as those given above, are irrelevant because the frequency of heterozygotes was already high. Wright further observed that a gene's specific function and its modifier activity on other loci can hardly be independent. Selection may tend to increase the frequency of an allele because of its modifying activities, but this change of frequency could well be opposed by its specific function. He suggests that dominance relates to differential activity between alternate alleles at a given locus. This explanation avoids the complexities of loci interaction and frequencies but it fails to take account of the fact that selection can modify dominance.

The principal problem with Fisher's initial explanation consists of the existing wildtype allele developing dominance over mutants as each appears. His process places heterozygote frequencies so low that the effectiveness of selection is questionable. An escape from this dilemma is possible if the wildtype alleles now observed were not at a selective disadvantage while at low frequencies. As selection favors their expression and their frequencies increase, the ratio of heterozygote to homozygote for the mutant exceeds 1.0 for possibly a long time, even after the gene is at an allele frequency of 0.5. Thus, the problem of low heterozygote frequencies no longer exists. The situation would be comparable to the evolution of dominance during a transient polymorphism. This operation suggests that alleles recessive to the existing wildtype were present during the time when it was initially climbing in frequency. Oscillations in frequency of an allele may accompany successive intervals when additional alleles become recessive, and one set of modifiers may also establish conditions for dominance over several similar alleles. Most observed mutations have phenotypes of low fitness, and the explanation suggests that the overhaul of dominance relations is not a frequent process.

Stable polymorphisms where a gene has pleiotrophic expression could also solve the problem of heterozygote frequencies. Namely, a balancing selection operating on one character would retain the heterozygote in high frequency. A change of phenotype proportions by dominance would not be expected in this character, but a second character determined by the allele could undergo dominance modification. Sheppard (1958) and Clarke (1964) also suggest how balanced polymorphisms and frequency dependent selection can explain the observation of a few dominant alleles that are rare and hardly candidates for a wildtype terminology. The feature of a higher frequency for the dominant allele is not a property of genes in a polygenic series controlling quantitative traits. This point is shown by Fisher (1930a, 1958) and discussed briefly below, but the conclusion predicts low frequencies for dominant alleles in a polygenic series and an equal likelihood for the dominance to increase or decrease the character's quantity.

The release of variability on which selection may act is closely tied to number of, and recombination between, chromosomes. The ability of chromosome number to change by selection is well reviewed for *Drosophila* by Patterson and Stone (1952). The discovery that crossover frequency was under genetic control and responsive to selection is also not surprising. An increasing body of literature exists revealing that both number and localization of chiasmata are under genetic control (Rees, 1955; Kidwell, 1972). This material is reviewed by Mather (1973) since it influences the recombination index (Darlington, 1939). This index is the haploid number of chromosomes plus the mean number of chiasmata per nucleus, and it gives a measure of potential recombination.

PROTEINS AND SELECTION

The initial gene action codes amino acid sequence in the polypeptide chains of proteins, and differences in these molecular patterns of closely related proteins presumably give a statement of accumulated evolutionary differences directly in the genes. A comparison of functionally similar proteins in a group of taxa, presumed to represent an evolutionary lineage, reveals the accumulated changes, and some workers have attempted to distinguish between selection and drift as the mode of operation accounting for the observations. Where time of separation for different lines can be judged, the rates of different amino acid substitutions can be calculated. The values of substitution rates are found to be higher than most selection values would predict; however, the investigators theorize that the rates are in good accord for a random nonadaptive substitution (Lewontin, 1974). This approach gets closer to the genetic material than a paleontologist does with fossil bones. While not fitting a conventional selection model, the data are not sufficiently compelling for the rejection of selection. In addition, occasionally those who adhere to a neutral explanation say the rates are constant. The so-called constancy appears to refer to an average for which no real constant property may exist. Lewontin (1974) rejects this feature of the neutral argument in view of the large variations in rates of evolution suggested on fossil evidence by Simpson et al. (1944).

Some amino acids have rather similar structures that also give them a similarity in function. If the substitution of amino acids by others with similar physiochemical properties occurs more frequently than predicted by chance, then a case for selection exists in the amino acid data. Clarke (1970) and others have found just such a pattern in the amino acid substitutions. Another approach by biochemistry-oriented investigators divides actual proteins into an essential group and a "useless" group. If selection controls the change of these molecules, the "useless" groups should be stagnant and remain static over long periods in evolutionary lineages. Should random drift be important, then the "useless" category could be expected to change rapidly with little correlation to pattern between related groups. Lewontin (1974) reviews the results, and the "useless" group appears to have changed more rapidly than selection predicts. However, as Lewontin points out, the argument is after the fact, and it is perhaps risky to assume that a given protein has the same level of "uselessness" in different species encountering quite different ecologies.

Chapter 4

The Balanced Polymorphism, or the Non-Neutral Equilibria

A stability of gene frequencies is not testimony for an absence of selection. Rather, equilibria of frequencies, though not Hardy-Weinberg values, are usually assumed to reflect genetic polymorphism in which gene frequencies oscillate about a fitness-determined balance. This conclusion comes from the theory of genetic polymorphism. This theory developed largely with R. A. Fisher, E. B. Ford, and their colleagues in England. Important implications for natural selection, not initially recognized, have emerged from the theory. These properties are discussed below; however, the early rationale is still judged valid. Genetically determined, discontinuous, alternative traits, i.e., polymorphic traits, are frequently found to co-exist in the same population and remain at fairly equal frequencies over successive generations. The stability of such frequencies is not an idealized operation of Hardy-Weinberg principles, i.e., a neutral equilibrium, if Fisher's (1930b) analysis is accepted. Fisher demonstrated that a factor maintaining a neutral selective value was exhibiting an exceedingly fine balance between advantageous and disadvantageous realms on either side. The likelihood of such a neutral value persisting for any reasonable time in the variable conditions of nature is negligible, and genes must either survive or vanish as their relations with selection dictate. Recent objections

103

to this view have developed with the recognition of so-called neutral alleles. The conflict appears to involve different workers recognizing different units or targets of selection. The question is discussed later, and Ford (1975), Dobzhansky (1970) and others are followed in assuming Fisher's conclusions correct. Certainly, the well studied cases of character stability have found selection operating. The examples of sickle cell anemia in man and numerous insect polymorphisms testify to this statement (Allison, 1964; Richards, 1961; Ford, 1975). If no escape from a selection scheme exists, we must expect the disadvantageous morph and its determining genes to decline in frequency until maintained only by mutational, or possibly migrational, input. Observed cases of polymorphism are abundant, but actual data on their stability exist for relatively few species. Consequently, one might assume without such data that morphs are in the process of substitution or transient polymorphism. The large number of simultaneously observed cases in different species is simply too great for this explanation to be valid for many cases. In fact, finding a substitution occurring in a natural population is the exception. Since elimination of characters fails to occur in most known polymorphisms and the absence of selection is unacceptable, only some form of balanced genetic polymorphism remains. Ford (1975 and earlier) defines it as ". . . the occurrence together in the same locality of two or more discontinuous forms of a species in such proportions that the rarest of them cannot be maintained merely by recurrent mutation."

Where discontinuous, polymorphic-type variation has been studied, it is almost invariably found to be genetically controlled. Consequently, Ford suggests that any case of discontinuous, alternative traits existing together in a population indicates selection. The discontinuous expressions for characters determined by major genes claim most attention; however, the process also applies for modifying-type genes if the same fitness pattern exists. If relative fitnesses of three genotypes at a locus are $AA >$ $Aa < aa$, a non-zero frequency of both A and a exists that may be described as an equilibrium. However, any displacement from these values leads to a progressively greater displacement. The result leads to fixation of one allele and the "equilibrium" is therefore unstable. The variable properties of natural populations clearly reduce the likelihood of this explanation for natural variation to the vanishing point. The genetic polymorphism model generally adopted recognizes a net, maximal fitness for the heterozygote, namely a situation where the before-selection genotypes, such as AA, Aa, and aa, experience fitness values of $1-s_1$, 1, and $1-s_2$, respectively. The Δq expression becomes zero at the stability or equilibrium point, and gene frequencies may be expressed in terms of the selection coefficients as given in Chapter Two. If the locus has a multiple allelic series, the possibilities become more complex. If alleles A_1, A_2, and A_3 exist, the heterozygous combination of A_1 with A_2, A_1A_2, may be less fit than A_2A_2, while A_1A_1 has an even lower fitness. However, the heterozygous combination of A_1 with the third allele, A_1A_3, may be sufficiently advantageous to overcome the losses. These possibilities are expanded more fully by Li (1967).

In recent years, the unit of selection discussed below has been increasingly equated to a large series of linked loci, and doubt has fallen on the importance of single locus heterosis, which non-neutral equilibria generally imply. The number of proven examples for single locus heterosis is not overwhelming, and Lewontin (1974) states that evidence exists for only one locus, that which controls sickle cell anemia in man. If one requires that the expression scored (for instance, red blood cell condition) must be shown to depend on alternate structures of a polypeptide produced by one segregating locus, then the case for single locus heterosis does stand on very little data. This state of affairs may well be due to the technical difficulties of collecting such data. On the other hand, selection has conventionally been assumed to operate on variation existing in the phenotype, and much variation is thus far removed from an initial gene product. If such variation is nonetheless transmitted in a single locus fashion, a demonstration of its heterosis supports the non-neutral equilibrium concept. The number of separate coding units in each segregating factor remains unknown, but the single-factor principle of Mendelian inheritance remains valid. With this less-restrictive view of single locus properties, the evidence becomes more abundant. An infrequently mentioned example began with Howard's (1962) summary of his own breeding data and that of his French colleague, Vandel, for the isopod *Armadillidium vulgare*. One trait discussed concerned red body color, which is dominant over the more typical grey color. The breeding tests for a large number of crosses and progeny clearly conformed to a single locus control in which *RR* and *Rr* genotypes have red bodies and *rr* is grey. Howard noted that the test cross results consisted of close 1:1 ratios for red to grey but the ratios of F_2 crosses came close to significantly differing from 3 red: 1 grey, which suggested a deficiency of the dominant phenotype, resulting perhaps from a loss of *RR* genotypes. Haldane (1962) subjected Howard's data to a further analysis, as follows. The numbers of offspring per class for test crosses and F_2 crosses are given as *a, b*, etc., such as:

Offspring Numbers per Class

Crosses	$(RR + Rr)$	Rr	rr
Test crosses		a	b
F_2 crosses	c		d

The ratio of Rr/rr in the test cross progeny is a/b, or y, which estimates the relative viability of Rr to rr for such sibships. The ratio of dominant phenotypes to recessives in the F_2 progeny is c/d. This value is ideally $(RR + 2Rr)/rr$. The relative viability of the Rr genotypes to the rr genotypes in the F_2 sibships may be assumed equal to the value for the same genotypes from test cross sibships. Thus $2y = 2Rr/rr$. The relative viability of RR to rr, given as x, may then be expressed as:

$$x = c/d - 2y.$$

Both y and x should equal unity if viabilities of the genotypes are equal; therefore, confidence intervals for y and x are required. Simpson, Roe, and Lewontin (1974) give a procedure for obtaining the variance of a ratio when the numerator and denominator are independent. The segregating numbers in these progeny are, of course, not independent, but an increase of numbers in one cannot be accompanied by a decrease in numbers of the other, i.e., they will vary in the same direction. Under this condition, the true confidence interval is less than the calculated value and the confidence statement becomes conservative, a desirable feature. Haldane gives variance expressions directly in terms of sample numbers, a, b, etc. that result in very similar confidence intervals to those obtained using the method of Simpson et al. Variances for a/b, or y, and c/d and $2y$ may thus be computed. The latter values are required to obtain the variance of x. The variance of x is the sum of variances for c/d and $2y$, because the variance of a difference equals the sum of variances for the two terms.

The value of y was found to be significantly greater than unity at the 5 percent level of confidence, and likewise, x was less than unity. The higher relative viability of Rr to rr was taken from test cross data in which the results could be due to gametic selection against r gametes in the heterozygous parent. However, this possibility is not supported by the F_2 results in which RR is indicated to be deficient. The selection appears to operate against diploids at some stage before scoring. The relative viabilities of the three genotypes at the R locus are thus $Rr > rr > RR$, and heterosis for factors segregating as a single Mendelian locus is established. It is interesting to note that one first uses the segregation data to identify the genetic mode of inheritance by its insignificant deviation from the single locus expectations. Then, the deviations are used to question the likelihood of unequal viabilities. The latter analysis requires considerably larger samples.

Where random drift of gene frequencies is likely, as in small populations, the relative magnitudes of the selection coefficients assume an importance. If the equilibrium point of gene frequencies is much displaced from 0.5, the polymorphism is at risk of being lost, as shown by Robertson (1962). Namely, if the equilibrium values are $\hat{p} = 0.8$, $\hat{q} = 0.2$, a chance-induced shift of p upward 0.2 above the equilibrium point results in fixation. If $\hat{p}$ is closer to 0.5, the same chance-induced event leads to no permanent effect. If s_1 and s_2, as used above, are largely unequal, then $\hat{p}$ and $\hat{q}$ will likewise differ, and the polymorphism is in jeopardy if drift is a possibility. This restriction of selection coefficients reduces the role that balanced polymorphism may play for small populations.

The magnitude of selection coefficients also assumes importance in a balanced polymorphism if inbreeding exists. The general effects of inbreeding are well discussed by Li (1955) and Mather (1973). Basically, inbreeding leads to homozygosity, and the rate of loss for heterozygotes is a function of the degree of inbreeding where self-fertilization is maximal. For purposes of illustrating how the magnitude of selection is involved, let m equal the percentage that are homozygotes and $1 - m$, or n, equal the percentage that are heterozygotes at a single locus. Let fitness be $1 - s$ for homozygotes and $1 - 0$ for heterozygotes and self-fertilization occur. Recalling that offspring from heterozygotes consist of half homozygotes and half heterozygotes, then one generation later the homozygotes will be m', or:

$$m(1 - s) + 1/2(n) = m'.$$

The total may be expressed as $m(1 - s) + 1/2(n) + 1/2(n)$, which reduces to $1 - ms$. Thus, the new m adjusted to a total of one will be:

$$\frac{m(1 - s) + 1/2(n)}{1 - ms}.$$

For equilibrium to develop, the following identity must exist:

$$\frac{m(1 - s) + 1/2(n)}{1 - ms} = m.$$

This expression can be algebraically revised as follows:

$$m - ms + 1/2(n) = m - m^2 s$$

$$1/2(n) = ms - m^2 s = ms(1 - m) = mns.$$

If $n = 0$, the expression balances but has no importance for polymorphism. If n is not zero, the expression becomes:

$$ms = 1/2,$$

$$m = 0.5/s$$

If $s < 1/2$, fitness of the homozygotes is greater than half that of the heterozygote, but then m exceeds one. Namely, the inbreeding goes to completion for full homozygosity. If $s > 1/2$, the homozygotes' fitness is less than half that of the heterozygote and m exists between 1/2 and one. Thus, where s exceeds 0.5 in such a mating system, equilibrium develops. If self-fertilization is replaced by weaker forms of inbreeding, s may be smaller for an equilibrium. For full-sib and half-sib mating systems, the threshold s values are 0.24 and 0.19, respectively (Mather, 1973). Other effects producing balanced gene frequencies exist without resorting to intrinsic heterozygotic advantage (Li, 1967; Cook, 1971).

These possibilities include the following interactions:

1. A mutation-selection equilibrium. Mutation occurs with a forward rate when the normal or more common gene mutates into a mutant allele. This form is usually detrimental with a less-integrated adjustment to other systems. The back mutation rate converts the mutant into the normal allele. Excepting the uncommon mutator genes and unnatural exposures to mutagenic sources, the rates are low, about 10^{-5} or 10^{-6} per locus per generation. These rates are estimates of the forward rate since the rarity of mutants restricts ability to measure back mutation. Back mutation appears, however, to exist at lower rates (Mather, 1973). If only mutation rates are in question, a gene's frequency will, after a large number of generations, come to balance with these rates. If, as the limited data suggest, the forward rate to the mutant form is highest, then equilibrium will consist of the mutant, ill-suited genes in greatest abundance (Mather, 1973). This prediction is totally at variance with facts, and the general conclusion is that strong selection intervenes to balance the excess forward rate thus producing an equilibrium whereby the mutant form occurs at a low frequency, probably a function of the forward mutation rate. The low frequency of the mutant class, while a consequence of selection, is often not detected, and such polymorphisms are rarely known outside of human disorders.

2. A migration-selection equilibrium. Genes arriving by migration probably do not carry the same degree of fitness disadvantage as genes coming by way of mutation, and the rates are perhaps not as low as seen for mutation. Also, two other important differences exist. Mutation is a single locus event, actually applying only to one member of a diploid locus. The units of migration are also discrete, but they consist of individuals, and such an individual carries a gene for every locus in the gene pool. Secondly, rates for migration surely vary widely with time as population densities oscillate. Equilibria are possible, as with mutation; however, the variable nature of migration rates suggests that few cases of stable frequencies in nature could be explained this way.

3. Nonrandom mating equilibria. A phenotypic assortative mating (like X like) is much less effective in developing homozygosity than inbreeding (Mather, 1973 and others), but may channel genes contributing similar expressions into the same lineage, an effect not seen with inbreeding. Phenotypic disassortative mating (like X unlike) is nature's effort toward an outbreeding system, i.e., an increase of heterozygosity above that achieved with random mating. This effect is small and of dubious importance for real populations. Where more than one locus is involved in a trait's determination, the increase in heterozygosity barely exceeds the 50 percent value obtained as a maximum with random mating (Mather, 1973). Cook (1971) gives an algebraic rationale for expecting an equilibrium with disassortative mating if each individual mates once and the two types occur in equal frequencies. This condition is too restrictive to occur commonly, and changing the algebra to accommodate realistic possibilities converts the operation into frequency-dependent selection. An equilibrium is also possible if mating is disassortative and the AA genotype is lethal. With matings of only $Aa \times aa$, an equilibrium obtains and A is not lost.

4. Periods of selection. Selection may act differently, though in constant magnitude, while in operation. The differences involve changes in time or space. The following possibilities apply to temporal differences for a single generation. Recognizing two periods of selection, the following fitness patterns may develop:

	AA	Aa	aa
First Period of Selection	$1 - s_1$	1	1
Second Period of Selection	1	1	$1 - s_2$

The net fitness over the full generation sums to $1 - s_1$, 1, and $1 - s_2$ for AA, Aa, and aa, respectively, and leads to an equilibrium, just as direct heterozygote superiority does. Wallace (1968, 1970) refers to this form of equilibrium as marginal overdominance. Selection operating on the same genotype but differently in two periods is reflected in the following way:

	AA	Aa	aa
First Period of Selection	$1 - s_1$	1	1
Second Period of Selection	$1 - s_2$		

Now if s_1 and s_2 have positive and negative signs, an equilibrium may well result. The second period of selection can act only on the frequencies remaining after the first period. Since the Aa and aa genotypes are now reduced the second period, the proportion of AA lost during the first period can thus be regained. Note that if s_1 were negative in sign, the situation would be equivalent to selection acting against Aa and aa. To achieve balance with this two-period pattern of selection, three different s values would be required, as follows:

	AA	Aa	aa
First Period of Selection	1	$1 - s_1$	$1 - s_1'$
Second Period of Selection	$1 - s_2$	1	1

The net fitness would be $1 - s_2$, $1 - s_1$, and $1 - s_1'$ for AA, Aa, and aa, respectively, and a balance requires that both s_2 and s_1' exceed s_1. If selection acts only on AA at one period, and upon AA and Aa during a second period, a situation possible with pleiotropic gene action, an equilibrium is also possible. The situation is shown as follows:

	AA	Aa	aa
First Period of Selection	$1 - s_1$	1	1
Second Period of Selection	$1 - s_2$	$1 - s_2$	1

The net fitness for the full generation sums to $1 - s_1 - s_2$, $1 - s_2$, and 1 for AA, Aa, and aa, respectively. If s_1 is plus (+) and greater than zero, and s_2 is minus (−), the fitnesses become $(1 - s_1) + s_2$, $1 + s_2$, and 1, respectively. Thus, Aa again has the higher net fitness.

These examples concern periods of different selection within one generation. If different patterns of selection operate on consecutive generations, an equilibrium is possible in which the after-selection gene frequencies of the second generation equal the before-selection values of the first generation. A simple example of this operation is illustrated by considering a species with a spring and fall generation and selection operating as follows:

Spring Generation (p_o, q_o)	AA	Aa	aa
Before-Selection Frequencies	p_o^2	$2p_o q_o$	q_o^2
Fitness	1	1	$1 - s_1$
After-Selection Frequencies	p_o^2	$2p_o q_o$	$q_o^2(1 - s_1)$

$$\frac{p_o^2 + p_o p_o}{1 - q_o^2 s_1} = p_1$$

Fall Generation (p_1, q_1)	AA	Aa	aa
Before-Selection Frequencies	p_1^2	$2p_1 q_1$	q_1^2
Fitness	$1 - s_2$	1	1
After-Selection Frequencies	$p_1^2(1 - s_2)$	$2p_1 q_1$	q_1^2

$$\frac{p_1^2(1 - s_2) + p_1 q_1}{1 - p_1^2 s_2} = p_2$$

If $p_2 = p_0$ an equilibrium results. The necessary s_2 value, given p_0, q_0, and s_1 values, can be obtained by algebraically adjusting the p_2 expression for a solution of s_2 when $p_2 = p_0$. The desired term is:

$$\frac{p_1 - p_0}{p_1^2(1 - p_0)} = s_2.$$

If $p_0 = q_0 = 0.5$ and $s_1 = 0.5$ in the spring generation, the p_1 and q_1 values are 0.5714 and 0.4286, respectively, and s_2 is 0.4373. Haldane and Jayakar (1963) describe a more complicated case of such selection. Many organisms have spring and fall generations in which selection is probably quite different; however, the balance of gene frequencies described above may not be a common event. The value of selection derives only a generation following its operation, and, in the above system, the consequence of selection is seen to increase the frequency of the trait that is disadvantageous. Thus, each generation pays a high price for genes that produce such effects. A gene action adapted to the mean of environmental variation would possibly involve lower total loss. The modification of a gene's action by modifier genes could well achieve this end. In the above example, an s_1 value of 0.5 was required against a q_0^2 of 0.25, while an s_2 of 0.4373 served to return the p value back to 0.5. The p_1^2 was 0.3265. Therefore, the target of selection in the fall generation was greater than in the spring and required a smaller s value to achieve a Δq of the same magnitude observed for the spring generation. This property concerns selection effect, discussed more fully below.

Examples of selection differences through space involve the sexes and habitat differences. The effect of increasing heterozygote frequency by randomly mating sexes that differ in gene frequency was mentioned earlier (the Robertson Effect). If sexes do in fact differ in gene frequency in a random mating system, the difference may logically be assumed to arrive by selection acting differently on the sexes. The possibility for such selection seems reasonable. If different but constant selection patterns involve one sex with heterozygote advantage and the other with heterozygous disadvantage, a balanced polymorphism is also possible. This kind of selection between sexes is perhaps not common but its operation is symbolized in tabular form as follows:

	MALES			FEMALES		
Genotypes	AA	Aa	aa	AA	Aa	aa
Before-Selection Frequency	p_0^2	$2p_0q_0$	q_0^2	p_0^2	$2p_0q_0$	q_0^2
Fitness	$1 - s_m$	1	$1 - s_m$	1	$1 - s_f$	1

After-Selection Frequency	MALES			FEMALES		
	$p_0^2(1 - s_m)$	$2p_0q_0$	$q_0^2(1 - s_m)$	p_0^2	$2p_0q_0(1 - s_f)$	q_0^2

$$\frac{p_0^2(1 - s_m) + p_0q_0}{1 - p_0^2 s_m - q_0^2 s_m} = p_m \qquad \frac{p_0^2 + p_0q_0(1 - s_f)}{1 - 2p_0q_0 s_f} = p_f$$

$$1.0 - p_m = q_m \qquad\qquad\qquad 1.0 - p_f = q_f$$

If selection is constant when operating and other possibilities for a change of gene frequencies negligible, the gene frequencies will adjust to the s_m and s_f values so that:

$$(p_f + q_f)(p_m + q_m) = p_0^2 + 2p_0q_0 + q_0^2,$$

and a balanced polymorphism results. The restriction of the required fitness pattern and its constancy seem likely to reduce the importance such systems might play in nature. Nevertheless, cases of unequal fitness for the same genotype in the two sexes have been reported. Smith (1975) reports a case involving sexual selection in the butterfly *Danaus chrysippus.* Two morphs exist in the species, *chrysippus* and *dorippus,* and the sexual selection involves mating vigor, not female choice. The selection pattern almost consistently favors females with the *dorippus* morph, judged by composition of mated pairs. Selection varies in favor of morphs among the males according to season and frequency. The effect is that both morphs are favored, each in a separate sex, and Smith contends that the process explains the dimorphism's balance. Parsons (1975) suggests that fast mating speed is favored in males, and slow speed in females, of *Drosophila melanogaster.* The genetic nature of mating speed is surely more complex than is exhibited for most visible morphs, as in *Danaus*; however, the selection pattern is also suggestive of a balancing effect.

If selection differs by habitat variation, the population may well be proportioned (probably unevenly) over the various habitats. If the total population of breeders nonetheless converges randomly to mate, an equilibrium may develop if selection regimes differ in such a way that they give a net maximal fitness for the heterozygote. The requirement of random mating between representatives of different habitats is not unlikely for a number of species; for instance, many frogs gather at a common pond to mate, insects with aquatic larval stages emerge from different local habitats to meet in a common mating swarm, etc. The habitats may in fact be of short duration but still perpetuate genetic differences into the breeding group. If the after-selection gene frequencies for A and a are p and q and F_1 development is distributed over three habitats where x, y, and z are percentages of the population in the three habitats, the same gene frequencies may occur in the before-selection periods of consecutive generations, as suggested by Levene (1953). The following arrangement of fitnesses may lead to similar gene frequencies before and after periods of selection:

Site	Before-Selection Genotype Frequency	Fitness	After-Selection Genotype Frequency	Adjusted Freq.	Adjusted Frequency times Habitat Proportion
AA	p^2	1	p^2	D^1	$D^1 x$
(1.) Aa	$2pq$	$1 - s_1$	$2pq(1 - s_1)$	H^1	$H^1 x$
aa	q^2	$1 - s_2$	$q^2(1 - s_2)$	R^1	$R^1 x$
				1.0	x
AA	p^2	$1 - s_3$	$p^2(1 - s_3)$	D^2	$D^2 y$
(2.) Aa	$2pq$	1	$2pq$	H^2	$H^2 y$
aa	q^2	$1 - s_4$	$q^2(1 - s_4)$	R^2	$R^2 y$
				1.0	y
AA	p^2	$1 - s_5$	$p^2(1 - s_5)$	D^3	$D^3 z$
(3.) Aa	$2pq$	$1 - s_6$	$2pq(1 - s_6)$	H^3	$H^3 z$
aa	q^2	1	q^2	R^3	$R^3 z$
				1.0	z

$$D^T = D^1{}_x + D^2{}_y + D^3{}_z; H^T = H^1{}_x + H^2{}_y + H^3{}_z; R^T = R^1{}_x + R^2{}_y + R^3{}_z.$$

$D^T + H^T/2 = p$ and $R^T + H^T/2 = q$, the before-selection values of the next generation. While this avenue towards an equilibrium, marginal overdominance in space acknowledges ecological diversity, it takes into account neither proportional changes that most probably occur in the density within different habitats, nor the unlikely stability of the fitness values. The assumptions are therefore quite restricting, even for the species that otherwise fit the system.

5. Frequency-dependent selection and soft selection. Selection coefficients discussed above can be viewed as constant, i.e., independent of existing gene frequencies or population density. Where these population properties play a role, the terms frequency-dependent and soft selection apply. Equilibria may result from such selection but are discussed more fully in a following chapter. Frequency-dependent selection involves a rare morph having a proportionately higher fitness than when it exists at high frequencies, and soft selection is a result of competition that is turned on and off as population numbers oscillate. Other possibilities for equilibria are occasionally noted as a balance between migration and genetic drift but omitting selection.

Equilibria involving mutation, gametic-zygotic interaction, sexual differences and direct heterozygote superiority may result from fitness being an intrinsic property of the individual. Fitness derived in this fashion is perhaps more stable than environmental controls can offer. For this reason, many workers feel that the early view of direct heterozygote superiority for balanced polymorphism is still largely valid. Nonetheless, studies rarely identify the actual mechanisms of selection behind balanced polymorphism.

Polymorphism, caused by balancing selection, produces many suboptimal genotypes from the segregation of genes at each heterozygous locus, i.e., the homozygotes. If segregation at each heterozygous locus is independent of other such loci, and the loss from segregating suboptimal types has an independent impact on total fitness, then the number of loci with allelic variations, i.e., those loci capable of segregation, is low, or the selection coefficients per locus are very small. Otherwise, fecundity is not usually high enough to pay the cost of segregational loss, locus by locus, and have enough offspring left to perpetuate the stock. Available estimates of selection coefficients are very rough; however, a number of studies cited below indicate values well above 0.1 and 0.2. These values may well be far higher than the true coefficient mean value but they document the potential for selection to operate at such levels. Consider a species with 10,000 loci, a conservative estimate, and 10 percent of these loci, or 1,000, are heterozygous in each individual. If segregation at each of these loci reduces fitness by one percent, or $w = 0.99$, then overall fitness would be 0.99^{1000} or 0.432×10^{-4}. Such a low value is less than potential fecundity of many species. Williams (1975) suggests that some species are "high fecundity" types, such as elm trees, certain marine fish, etc., and can afford the high cost of independent contributions toward reducing fitness. However this suggestion is resolved, a great many species usually studied for evolutionary processes do not possess the reproductive output needed to overcome the predicted loss of even 10 percent of an individual's loci to segregation into a pattern of balanced selection with modestly valued coefficients. If selection coefficients are small enough to allow independent contribution to fitness loss by separate loci, then they are essentially neutral, since minor environmental variations would swamp their influences. While estimates of selection coefficients are lacking or of poor accuracy, the estimates of loci proportions in the heterozygous state are rather well developed. Lewontin (1974) reviews this material, concluding that 30–40 percent of loci in a population and 10 percent in an individual may be heterozygous. The tremendous potential for such systems to reduce fitness has refocused attention on the concept of neutral genes, sometimes called "genetic junk." The few cases in which clear intrinsic advantages of the heterozygote have been documented do not mean that all or most stable variation is maintained by a balancing selection. Actually, Lewontin's review points out that inbreeding depression is less than expected if existing heterozygosity plays an independent role in determining fitness. The question gives perspective on the actual unit of selection. The great majority of data discussed in balanced polymorphism do not specifically identify the genetic basis of expression for the unit. For instance, a single locus or a large block of linked loci are frequently difficult to distinguish between in segregational data, but are important in interpreting fitness. This aspect of selection is considered later; however, as with the earlier version of balanced polymorphism, if selection and stability simultaneously affect a trait, the conclusion of heterozygous advantage is difficult to escape.

Numerically documenting a balanced polymorphism regardless of the balancing mechanism can be complex when both before- and after-selection samples are not available. Samples within one generation having equal p and q values but different genotype frequencies would identify a balanced polymorphism involving differential survival. A sample of breeders of one generation, paired with a sample of their young, composes a comparable basis for recognizing differential reproduction. Unfortunately, such data are typically difficult to acquire. Use of single-sample, after-selection data requires the assumption Fisher made for *Paratettix*, and some independent evidence is desirable on this point. Successive after-selection samples having the D, H, and R proportions equal but not in equilibrium with their p and q values (also equal) constitute evidence for a balanced polymorphism. However, successive before-selection samples with similar p and q values and genotype proportions may describe a neutral equilibrium. Characters expressed in external phenotypes are much easier to analyze, because individuals may be scored directly and they provide direct estimates of after-selection values. Note that the frequently quoted *Drosophila* data involve before-selection samples. Pair matings circumvent this problem, since scoring a sufficient number of offspring for chromosome composition allows a fairly good identification of the chromosome composition of the parents. In cages and in nature, however, pair matings cannot be assumed, and the actual after-selection frequencies remain unknown. Balanced polymorphism is judged present by observing events in population cages, an impossible technique for most species. Populations are started with known, selected inversion frequencies. Adults are removed periodically and allowed to oviposit, and the resulting larvae (before-selection) are scored. For a given cage condition (temperature, etc.) frequencies can be followed and observed to approach stable values for a given environment, although the same heterozygote appears to have maximal fitness over wide range of conditions. The data discussed by Dobzhansky (1961) illustrate this reasoning.

Chapter 5

Selection Coefficients in Natural Populations

Estimates of selection coefficients may be envisioned for full- and partial-generation intervals. Full-generation coefficients are, of course, desirable for the best prediction of evolutionary consequences of selection. These estimates would also be net values resulting from several selection agents and, as a consequence, not representative of the action of any one agent. Partial-generation coefficients representing relative survival during one stage of the life cycle, or the action of only one of several selection agents, may often be directly associated with a specific predator activity, etc.; however, such coefficients have less predictable potential on a character's evolutionary future. Full-generation coefficients are mainly difficult to obtain because one's actual samples rarely represent all selection (Prout, 1971a, b). The earliest estimate of a selection coefficient, as used here, was obtained by Haldane (1924). His basic expression for the coefficient of selection, k or K in his usage, was $\int_0^\infty kdx$, where x represents different age intervals. Namely, the net coefficient is the summation of coefficients operating at each age. This approach is largely academic since the necessary parameters are usually not available and beyond field methods to acquire. His alternate approach involved deriving a mathematical expression giving a numerical value of the selection coefficient per generation times the number of generations over which a measured change in gene frequency occurred. This method may be illustrated simply where selection operates against the recessive homozygote. If for generation n, the frequencies of A and a are p_n and q_n, or $1 - p_n$, respectively, then the following conditions exist:

Genotypes	AA	Aa	aa
Before-Selection Frequencies	p_n^2	$2p_n(1-p_n)$	$(1-p_n)^2$
Fitness	1	1	$1-s$
After-Selection Frequencies	p_n^2	$2p_n(1-p_n)$	$(1-p_n)^2(1-s)$

The ratio of A to a in n is $p_n/1 - p_n$, or U_n, and the relative proportion of $AA{:}Aa{:}aa$ in n after selection is $p_n^2 : 2p_n(1 - p_n) : (1 - p_n)^2(1 - s)$. From this genotype ratio, an expression for U_{n+1} is possible. If each term is divided by $(1 - p_n)^2$, we obtain another expression for the genotype frequencies, i.e.:

$$U_n^2\; AA :\; 2U_n\; Aa : (1 - s)\, aa.$$

The value of U_{n+1} is then obtained as follows:

$$U_{n+1} = \frac{2U_n^2 + 2U_n}{2\,U_n + 2\,(1 - s)} = \frac{2U_n\,(U_n + 1)}{2(U_n + 1 - s)} = \frac{U_n(U_n + 1)}{U_n + 1 - s}$$

This expression is a recurrence relationship, since it gives a series where each event depends on the result of the previous event. The events here are ratios of gene frequencies in each generation that in turn result from a given selection coefficient. If the generation number is large, the solution for U_{100}, for instance, would require considerable labor to compute. However, if s is small, 0.01 or less, then a large number of generations may be expected in which the change in a specific U is approximately equal to the change in the generation just before and after. Thus, if U is plotted on the y-axis and n on the x-axis, the slope can be given as:

$$\frac{U_{n+1} - U_n}{(n + 1) - n} = U_{n+1} - U_n.$$

Namely, the change of U_n with n can be expressed as:

$$\frac{dU_n}{dn} = U_{n+1} - U_n.$$

Substituting the above expression for U_{n+1}, we obtain:

$$\frac{dU_n}{dn} = \frac{sU_n}{U_n + 1 - s}.$$

Since the expression was obtained by assuming that the change of U with n was similar between successive generations due to a small s, a further use of the small s assumption is possible. The right-hand term of the equation reduces to approximately $sU_n/(U_n + 1)$. From this result, the equation can be expressed as:

$$(U_n + 1)dU_n = (sU_n)dn, \text{ or } (U_n + 1)dU_n \cdot \frac{1}{U_n} = sdm, \text{ and}$$

$$\int_0^n (1 + \frac{1}{U_n})dU_n = \int_0^n sdn; \; U_n - U_0 + \log_e U_n - \log_e U_0 = sn, \text{ or}$$

$$U_n - U_0 + \log_e U_n/U_0 = sn.$$

In this way, the number of generations required for a given change under a constant selection coefficient could be predicted, or if the genetic change and generation number were known, the selection coefficient could be estimated. When s is not small, rapid selection in Haldane's usage (1932), Haldane derives an alternate but more complex solution that still treats data covering a number of generations. The likelihood of a selection coefficient remaining constant over a large number of generations seems low, and Haldane's answer may not be very good for any particular generation. He applied these methods to the melanism found in the now well-studied species, *Biston betularia*. The melanic morph is the expression of a dominant allele at a single gene locus, and the activity of lepidoptera collectors in England has left a rather good history of this morph's change through time. In Manchester, the first recorded specimens of *carbonaria* were taken in 1848, and 47 years later, in 1895, 98 percent of all Manchester specimens were melanics (Ford, 1975 and earlier). The species has one generation a year in the Manchester area. Haldane's calculation gave the coefficient acting against the typical morph as approximately 0.33, or only two of three typicals achieving the same success as each of three *carbonaria*. This effect is equivalent to a 50 percent advantage to *carbonaria*. In Haldane's words, ". . . the fertility of the dominants must be 50% greater than that of the recessives," and, he added, ". . . not very intense degree of natural selection." This value was much larger than most students of selection were prepared to accept in 1924. Subsequent studies, described below, have reached essentially the same numerical value by very different methods. The extent to which the estimates are comparable is not clear. Haldane's value is a mean over several generations that presumably reflects both survival and "fertility." Later workers have been concerned with only differential survival. Haldane (1942) also applied this approach to the study of selective disadvantage experienced by silver foxes in Canada due to the fur industry. The silver attribute is due to a dominant allele, but here the dominant is being selected against, and his ratio consisted of recessive to dominant gene frequencies. Elton (1942) presented data collected from the records of fur collectors covering a period from the 1830's to the 1920's. He concluded that the fox populations were not endangered but that the fox hunters, and to a less extent, the trappers, were responsible for the progressive fall in frequencies of the silver fox in each year's take. The silver fox skins brought higher prices and were naturally sought more seriously by the hunters. To apply Haldane's calculation of a selection coefficient, several adjustments of unknown accuracy were necessary. The actual data represented the foxes lost to selection rather than the after-selection values, and the number of generations is also not clear. Likewise, the commercial production of silver foxes by raisers was assumed to have made no impact on the market by the 1920s. The results gave a selective disadvantage of 0.03 to 0.04 to silver foxes per generation over the 90 years preceding 1920. These estimates, while subject to considerable error, represent the early efforts to measure directional selection.

The study of *Paratettix* by Fisher involved samples that were assumed to reflect all selection. Recall that Fisher used a phenotypic proportion form of gene frequency analysis rather than the cross-product ratio. He found the observable double dominants (where he equated Ab/aB to A/B) were only 60 percent of expected numbers. Thus w for the double dominant class in generation was 0.6, and s was 0.4. At least 13 dominant, linked genes were judged to be operating in the polymorphism, and the expected frequencies for the double dominant combinations ranged from 0.0825 to 0.2073; thus, only 60 percent of these values, or 0.0495 to 0.1244, presumably survived each generation. The case in which the sampling age is assumed to reflect all selection and migration of a generation has also been analyzed for a full-generation coefficient relative to geographic clines and is discussed below.

Sheppard's (1951) work with *Panaxia dominula* illustrates use of the cross-product ratio and a regression method for multigeneration data. Recall that Fisher and Ford (1947) presented evidence earlier that gene frequency changes for a population of this moth were due to selection. Sheppard's data treat the same population but incorporate four additional generations. The moth exists in three phenotypes, *dominula*, p^2; *medionigra*, $2pq$; and *bimacula*, q^2. The common form, *dominula*, judging from several years of data, possessed maximal fitness and the objective was to obtain estimates and an explanation of *medionigra*'s fitness. The sample was made of adults, assumed to represent after-selection values. Gene frequencies of these samples gave, therefore, genotype frequencies for the next generation in the absence of selection. An estimate of *medionigra*'s fitness for each of 12 years was obtained from the expression:

$$\frac{p^2}{2pqw} = \frac{\text{number of } dominula}{\text{number of } medionigra}.$$

The value of w was observed to vary between 0.59 to 1.42 but q fell from 0.11 in 1940 rather regularly to 0.04 in 1950. Sheppard assumed this decline resulted from a constant $w < 1.0$ for $2pq$, and the fitness of q^2 could be expressed as function of the w for $2pq$. The latter assumption simplifies the analysis. The frequency, q^2, of *bimacula* was always low (3 specimens or less in samples exceeding 200). Consequently, q^2 made little influence on the data and its fitness could vary largely without affecting calculations; recall the small Δq for low q values at even large s values. Sheppard considered the fitness of *bimacula* therefore as w^2. With these assumptions, the value of p in the next generation is:

$$\frac{p^2 + pqw}{p^2 + 2pqw + q^2 w^2} = \frac{p}{p + qw}.$$

Sheppard calculated a regression line for q on successive generations expressing q values as $\log_{10}$ values on the y axis and with years (1939 to 1950, or 0 to 11) on the x axis. Since the relation of Δq to q is curvilinear, large differences in Δq may occur for a constant s value. The log transformation corrects this feature. For a short region of the q range, the regression of q on successive generations may thus approximate a straight line, particularly for small s values. The equation for the line then gives the log of the expected q per year. Taking successive estimates of the expected gene frequencies and placing these values into the above expression gives an estimate of the constant w inherent to the data. Sheppard used the maximum likelihood method for this regression line, obtaining an estimate of 0.907 for the constant w. This method is considered most accurate, but if one recalculates the regression line by the more commonly used least squares method, the value obtained for w is 0.898 or a difference of 0.009, hardly enough here to affect interpretations. The coefficient acting against *medionigra*'s fitness is 0.093 or 0.102, by the two methods respectively.

The validity of using a constant w may be judged by a χ^2 test comparing expected and observed numbers of the recessive (or *medionigra*) gene per sample. Sheppard reports a χ^2 of 1.51 with 10 degrees of freedom (11 years of data), thus a $P > 0.10$ indicating a constant w sufficiently explains changes in the data. If variances for single-generation estimates of w, taken from cross-product ratios, were computed for the data, w's for the most years would probably not significantly differ from one. The low frequency of *bimacula*, q^2, expedited this analysis; however, higher frequencies of this morph would require more accurate estimates of its fitness in terms of the w for $2pq$.

Wright and Dobzhansky's (1946) treatment of *Drosophila* data, discussed above, also assumed constant fitness. Sheppard's work concerns a population experiencing variations in a natural habitat, while the *Drosophila* lived in a controlled laboratory where constancy may be a more valid assumption. The *Drosophila* study concerned a balanced polymorphism of chromosome inversions and required an estimate of two simultaneously acting coefficients. The Δq method utilizing $\overline{W}$, described in Chapter Two, was used. They found t and s values of 0.289 and 0.680 in a fitness set of $p^2(1 - t)$, $2pq$, and $q^2(1 - s)$. The relation of these values to the cross-product ratio was also given in Chapter Two, but these authors note that more accurate estimates are possible. They point out that for constant values, the s and t values minimizing the squared deviations of observed and estimated Δq values are the best estimates. The effect is somewhat like Sheppard's regression method. To achieve this goal, they utilize partial derivatives in an iterative (trial and error) method. The solution they accept gives 0.304 and 0.695 for t and s, respectively. An increase of 0.015 occurs for each coefficient, or an increase of 0.05 and 0.02 percent of t and s, respectively, suggesting that adjustments may be more valuable for smaller coefficients.

A rather different approach exists for estimating full-generation coefficients if variation takes the form of a cline (Haldane, 1948). Clines adjusted directly to environmental selection agents are concerned. When a character's frequency changes from a low to a high value over a distance axis, it is logical that the character's relative fitness is low in one direction and high in the other. In fact, it is possible to picture this change of fitness as occurring at a sharply defined boundary, as described in an earlier chapter, while the curve of the cline passes in a smooth slope through the boundary, owing to the effects of migration. At the point, fitness is considered as 1.0; it changes to greater or smaller values to either side. Haldane assumed that approximately equal densities of animals (or plants) mating randomly occur along the distance axis in addition to other assumptions mentioned below. These criteria are not unreasonable and, as Haldane indicates, may deviate somewhat without invalidating conclusions on selection. The theory is developed with rather difficult mathematics but can be verbally explained as follows.

The cline records one phenotype (or allele) from a region where it has high frequencies to one where it has low values. The character may also be a quantitative expression by classing variation into large and small pheno-types, etc. The frequency of a character or class is plotted by percentages relative to the geographic axis. A sigmoid shape may represent the curve that simply observes frequencies changing faster per unit distance in some regions than in others, a not uncommon observation. The slope and geographic position of the cline are assumed to be stable, and migration across any given point on the cline, excepting the extreme ends, is assumed to be equal in areas of lower and higher frequencies. If the change of character frequency with change of distance is approximately equal on either side of the fitness boundary, then absolute magnitude of the selection coefficients should be approximately equal. Such a fitness boundary would occur near the middle of the sigmoid curve or near the inflection point. Stability of a cline may also develop if the selection coefficients to either side of the fitness boundary are unequal. In this case, migration, still assumed equal in terms of individual numbers, adjusts the position of the curve. The lower selection coefficient becomes more effective in reducing the frequency of its target as the frequency of that target increases (see discussion of selection effect below). Simultaneously, as the percentage of the target for the weaker selection coefficient in-creases, the percentage of the phenotype adversely affected by the higher coefficient decreases. Depending on the difference between the two coeffi-cients, a curve will eventually stabilize in relation to the fitness boundary. The equilibrium results when the weaker coefficient operates against an abundant target and cancels the migrant effect on one side of the bound-ary while the higher coefficient operating against a less-frequent target cancels migrant effect on the opposite side. The equilibrium frequencies are thus determined by the selection coefficients. A cline stabilized about a fitness boundary separating unequal selection coefficients gives an un-equal change of character frequency with change of distance to either side of the point. Such points are displaced toward either end of a cline, e.g., near or beyond the 25th and 75th percentile points. Clines adjusting to unequal selection coefficients would also shift geographically toward the region generating the weaker coefficient. The situation may be graphically illustrated by plotting the frequency of genotype, q_2, as in Figure 1. To the left, q^2 is reduced by coefficient s_1, and to the right, is increased by s_2 (thus using a fitness greater than one for comparative purposes). If curve 2 depicts the case where $s_1 = s_2$, the fitness boundary would be near b. If curve 1 reflects the case where $s_2 > s_1$ and curve 3 where $s_1 > s_2$, the curves are each shifted toward the area of the weaker coefficient. Yet, with appropriately chosen coefficient values, b could be the site of the fitness boundary for all three curves.

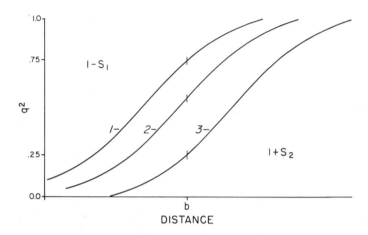

Figure 1. Three possible positions for a cline relative to the fitness boundary, b. At b, fitness of $q^2 = 1.0$; to the right and left, fitness is $>$ and $<$ 1.0, respectively. Curve 2 places b within the interquartile range of q^2, while b lies to the right and left of the range in curves 1 and 3, respectively.

The parameters required in obtaining Haldane's solutions for the selection coefficients are the cline's slope and geographic position, the fitness boundary's location, and an estimate of migration. The boundary, if such a region as was defined exists, must be located independently, but some guidelines are suggested. If the character scored is that determined by a recessive gene at one locus, then Haldane shows a functional relationship between the coefficients and the frequency of the recessive gene on the boundary. In his Table 1, he gives selected s_2/s_1 values for corresponding boundary q values covering a q range from 0.1 to 0.9. Haldane uses the symbols k and K for s_1 and s_2, respectively. The lower and upper quartile values are: for $q = 0.25$, $s_2/s_1 = 0.0417$ (by interpolation); and for $q = 0.75$, $s_2/s_1 = 2.821$. If the boundary exists at or below the lower quartile point (approximately curve 3), then $s_2 = 0.042s_1$. If the boundary exists at the upper quartile point (approximately curve 1), then $s_2 = 2.82s_1$. These coefficient functions refer to a single locus; however, the general conclusion applies to other cases. Namely, as the frequency of alternate phenotypes becomes more unequal across the boundary, the more unequal the coefficients become to either side of the fitness boundary. Such differences in net selection coefficients to either side of a fitness boundary, as defined, seem very unlikely. It might be added that while distinctly different coefficient values, differing both in magnitude and sign, are unlikely across a sharp boundary, equal but large coefficients, differing only in sign, are also unlikely. A boundary separating rather small coefficients located within the interquartile range, d, is reasonable, and this range can be estimated from data plotted on the distance axis. The estimate of mean migration, movement from birth to breeding, per generation must be obtained by population sampling procedures suitable for the species concerned. Migration expressed in this fashion is symbolized by m.

Reasoning along these lines, Haldane derived estimates for s_1 and s_2 in terms of m, d, and certain constants. His expressions are: $s_1 = (1.27m)^2/d^2$ or $(0.59m)^2/d^2$ for q and q^2 data, respectively, if d lies completely to the left of the boundary (as in curve 1); or $s_2 = (0.79m)^2/d^2$ or $(0.811m)^2/d^2$ for q and q^2 data, respectively, if d lies completely to the right of the boundary (as in curve 3). The expressions give only estimates for either s_1 or s_2 if the boundary is not within d; however, the magnitude of the other coefficient could be judged from Haldane's table mentioned above. Large differences in s_1 and s_2 are unrealistic, as suggested, indicating the boundary to be within d. The mean s_1 and s_2 in this interval is $(1.03m)^2/d^2$ or $(0.70m)^2/d^2$ for q and q^2 data, respectively. The q^2 values refer to phenotypes scored. Presumably, if phenotypic variation is simply classed into two categories, large and small for example, one could apply similar reasoning to their frequencies.

Haldane applied these expressions to Sumner's study, discussed above, of *Peromyscus polionotus* in northwest Florida. He estimated d as 12 miles, judged m from other *Peromyscus* work to be 0.5 miles, and considered the boundary most likely to lie within d. The region of the curve within d having maximum frequency change per distance, namely near the inflection point, has a q^2 value of 0.54. The average value of s_1 and s_2 within d is $[(0.70)(0.5)]^2/12^2$. The value rounds to 0.001, a rather small coefficient. Documenting a coefficient of this size by Δq or cross-product methods would be nearly impossible, as mentioned above. Haldane notes, "If the calculation made above is even roughly correct, ... selection could only be detected with certainty by observations on tens of thousands of animals." Selander et al. (1971) have suggested that population subdivision near the coast may account for the cline in part; however, Haldane seems to have considered this feature with ". . . a barrier which is difficult to pass would be the equivalent of an increased distance."

Examination of the expressions for s_1 and s_2 show that large m values are required relative to the size of d to obtain larger coefficient values. Haldane's model thus relies heavily on migration to explain a cline's slope. If each point on the distance axis has its own fitness regime rather than selection changing at only one boundary, then the process of compressing coefficients to the left and right of a point results in averaging many high and low values. Such an operation would give a small value, as observed, unless the point were well displaced from the curve's midway region, where one coefficient would be less affected. The actual magnitude of selection may thus be greatly underestimated by the mathematical simplification. Bishop (1972) reports a cline study, discussed below, in which each point is in fact found to have its own fitness conditions. This limitation is mentioned again relative to the location of the fitness boundary.

Several sets of cline data have been analyzed with Haldane's expressions; perhaps the most extensive study being that of Kettlewell and Berry (1961). These authors studied a nonindustrial melanism in the moth *Amathes glareosa* in the Shetland Islands. A melanic form, *edda*, replaces the typical moth at the northern end of a north-south cline through the islands. The melanic morph appears to be determined by a single gene pair where the heterozygote may, with experience, be recognized. Data for 22 sites along the cline exist, and a plot of the typical gene's frequency with distance appears in Figure 2, accompanied by a free-hand approximating curve. The north-south axis is reversed from a graph in Kettlewell and Berry's paper, to conform with curves in Figure 1 above. The cluster of points at the site 67 miles south of the northern end of the islands corresponds to a q value of 0.90 and occurs in the Tingwall Valley. This region represents the most obvious ecological change in the islands and the frequency of the typical gene falls progressively north of this area. Kettlewell and Berry were impressed with these factors and identified this valley as the fitness boundary. The boundary was thus positioned by ecological conditions rather than slope of the curve, and consequently d lies completely to the boundary's left. They analyzed their data accordingly, taking m and d as 0.25 and 48 miles, respectively. They obtained an s_1 value (k in their paper) of 0.00004. If q at the boundary equals 0.90, for which Haldane's table shows s_2/s_1 as 18.49, then s_2 would equal 0.00074. Kettlewell (1961) was in fact unable to obtain significant differences in survival studies of the two morphs, thus also pointing toward low or nonexisting coefficients. Later, Kettlewell and Berry (1969) revised their estimate of m to 0.5, thus changing their estimate of s_1 to 0.002 and of s_2 to 0.004, values still too low to verify statistically. Actually, no reason exists to believe the Tingwall Valley is a fitness boundary, in Haldane's sense, on the basis of the curve's slope. South of the valley, no cline seems to exist and because the valley is known to restrict migration (Kettlewell and Berry, 1969), the data from the valley northward represents the best approximation to Haldane's cline model. Using this view, d is about 25 miles and s_1 or s_2 within d equal $[(1.03)(0.5)]^2/25^2$, or only 0.0004.

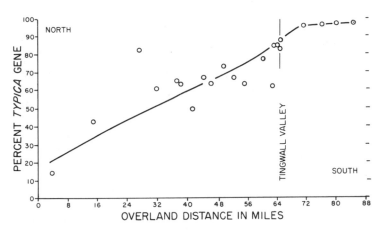

Figure 2. Clinal changes in gene frequency of the moth *Amathes glareosa* in the Shetland Islands. Revised from Kettlewell & Berry (1961).

While survival data and the cline model analysis both point to small selection coefficients, the experience of Kettlewell and Berry (1969) in the field convince them nonetheless that strong selection operates.

An examination of Haldane's expressions reveals that the model becomes unrealistic with coefficients approaching or exceeding 0.1. Clearly, m is not likely to equal or exceed d or even equal $0.5d$; therefore, for all reasonable relations of m and d, the coefficients are quite small. If actual net coefficients are of a magnitude of 0.1 or greater, Haldane's concept of a fitness boundary must be invalid, as suggested above. On the other hand, if the coefficients are indeed small, then the argument for locating the fitness boundary within d is less compelling. For instance, if s_1 is 0.2, then at the lower limit of d, s_2/s_1 is 0.042 and s_2 becomes 0.0084. For 500 individuals to either side of the boundary at $q = 0.25$, 100 and 4.2 of $(1 - q^2)$ and q^2 classes, respectively, will be eliminated. This differential seems unlikely across a sharp boundary in nature. If s_1 is 0.02, s_2 becomes 0.00084 at the same boundary, then only 10 and 0.42 of the comparable classes will be eliminated from 500 of each. This differential, while of the same relative magnitude, is more plausible; therefore, Kettlewell and Berry's description of the Tingwall Valley as a boundary is possible. The highest coefficients apparently reported by using Haldane's analysis are 0.055 and 0.005 for plant height and leaf length in *Anthoxanthum odoratum* (Antonovics and Bradshaw, 1970).

Partial-generation coefficients, involving of course within-generation data, are adaptable to analysis by cross-product ratios. However, the earlier estimates of selection for such intervals did not use this approach. The

selective disadvantages were identified by: (1) comparing morph proportions in recapture or predated samples to expected proportions judged
from numbers previously marked and released, or (2) comparing partial
absolute fitnesses also computed from observed and expected numbers in
recapture samples.

These studies were largely initiated by Kettlewell's work in industrial
melanism in the moth *Biston betularia.* His results initially appeared in
1955 with a report on capture-mark-release-recapture data about Birmingham, England, an area affected by the fall-out of coal dust resulting from
the local industries. He found 27.5 and 13.0 percents of released *carbonaria* (the melanic) and typical morphs, respectively, in the recaptures, a
ratio of about 2:1, thus confirming greater survival of the melanic form in
polluted areas. A year later, he reported additional tests and summarized
existing data (Kettlewell, 1956a, b). These reports include more Birmingham data and results from a non-polluted region, Deanend Wood, Dorset.
The results were again given as recapture proportions. For instance, in one
test, 154 *carbonaria* and 64 typicals were released in Birmingham with 82
and 16 recaptured, respectively. These results show 53.25 and 25.0 percents recaptured for *carbonaria* and typicals, respectively, again about 2:1.
A measure of selective disadvantage of the typical morph was obtained by
calculating expected numbers in the recapture sample. The values are
obtained by:

$$\text{expected recaptures of morph } A = \frac{\text{releases of morph } A}{\text{released total}} \text{ (recapture total)}.$$

Using these values in the ratio of observed/expected morph A recaptures
gives a component of absolute fitness as defined above. For example, in the
Birmingham region, the main data come from 630 released males including
137, or 21.7 percent, of the typicals. The recapture sample totaled 166, so
the expected number of typicals was 36.098 (Kettlewell tabulated recaptures from day to day using a death rate adjustment which gives a value of
35.9658 expected typicals). Only 18 typicals were observed to be recaptured, so the absolute fitness component is 0.4997 or 0.5. The coefficient
is clearly a partial-generation measure of absolute fitness; however, Haldane (1956) in a companion paper to Kettlewell's resumé, uses the 0.5
value as a relative s for a calculation assuming full-generation coefficients.
Actually, Kettlewell's data appears in sufficient detail for a cross-product
ratio estimate of partial relative fitness. Eighteen typical and 140 *carbonaria* morphs were recaptured (8 of the rarer morph, *insularia*, complete
the 166 total). His data give 35.9658 and 121.4615 expected numbers,

respectively, of these morphs assuming no selection. Solving for w in the following expression:

$$\frac{35.9658\,w}{121.4615} = \frac{18}{40}$$

gives 0.4344, a value somewhat lower than the 0.5 mentioned above. For the Deanend Wood, 67 typical and 32 *carbonaria* morphs were recaptured, while 54.2535 and 47.4180 were expected. Thus w, now the relative fitness for *carbonaria* to the typical morph, is 0.5465. The fitness of *carbonaria* in its less-suited habitat appears higher than that of the typical morph under similar conditions; however, the small recapture numbers suggest the difference is not significant.

Perhaps the most ambitious effort to measure selection was Kettlewell and Berry's (1969) study of survival among recaptures of the moth *Amathes glareosa* in the Shetland Islands. Recall the cline discussed above in which the melanic form *edda* occurs in high frequencies at the northern end of the islands and is progressively replaced southward by *typica*, the non-melanic form. Kettlewell suspected the selective agent was bird predation; however, the direct observation of a sufficient number of predatory acts was impossible to obtain. Also, the typical morph was so infrequent at the northern end of the cline that sufficient numbers could not be collected locally for marking and release. Consequently, he set up a program whereby samples of *typica* were collected in the southern region, marked, and moved by several means over a 24 hr period to the northern site, Unst, for release. Kettlewell also analyzed this data by a component of absolute fitness. He obtained 132 and 95 observed *edda* and *typica*, and computed expected numbers of 123.125 and 103.875, respectively. The ratio of 132/123.125 is 1.072, and he notes that *edda* appears to have approximately a 7 percent advantage. The comparable fitness for *typica* was 95/103.875, equaling 0.9145. Since these values are absolute components, the relative fitness of *edda* to *typica* would be 1.072/.9145 or 1.17. Of course, the direct cross-product ratio gives the same result:

$$\frac{123.125\,w}{103.875} = \frac{132}{95} \;; w = 1.17.$$

The comparable relative fitness of *typica* is 0.853. Thus, the relative fitness advantage of *edda* seems closer to 17 than 7 percent, again suggesting that the low coefficients obtained from the cline model are misrepresentations. Kettlewell's data did not reach significant differences for recapture proportions and perhaps relate closely to findings reported in 1969. He then noted that individuals released at points other than sites of original capture have lower survival. This fact could reduce below true values the estimate of selective disadvantage suggested for *typica*.

Cain and Currey (1968) studied selection in the snail *Cepaea nemoralis* by comparing data in samples of predated individuals. The snails exist in several morphs involving 0 to 5 pigment bands on shells having background colors of either yellow, pink, or brown. The predators (in this study) were Song Thrushes, *Turdus ericetorum*. These birds first locate a snail, carry it to a "thrush anvil," or stone, where the shell is broken, and then they eat the soft body. A suitable stone is visited repeatedly by local thrushes and their nature of breaking shells is sufficiently characteristic for identifying the predator as a thrush. The breakage is not so complete as to reduce a shell to fragments and, as a consequence, remains around an anvil can be sorted into the number of shells, their phenotypes scored, and marks on any shells recorded. By regularly collecting predated shells around the anvils, a chronological record of bird predation that reveals selectivity and relative predation pressure develops. The anvils are utilized by birds searching through adjacent but distinct snail habitats; i.e., areas where frequencies of the snail morphs differ. Therefore, the anvil collections must also be sorted for habitats in which the birds found them. To obtain these estimates, the authors determined morph frequencies of living populations from habitat-sorted samples. Samples from each recognized habitat were specifically marked and released. The recaptures gave estimates of proportions marked in each population. The marked shells appearing at anvils gave proportions of total predation per habitat. The predated but unmarked shells could then be assigned proportionately to the habitats, making the total anvil sample useful. The morph frequencies in the living population allowed a habitat's unmarked allotment to be partitioned into expected morph frequencies, and the frequencies observed in the predated, marked shells allowed the allotment to be partitioned into "observed" numbers. Thus, the total anvil sample may be sorted into habitats and expressed in expected and observed numbers. Mainly background colors were examined, and the following table summarizes data for a habitat where selection was most likely:

	Morphs		
	Yellows	Pinks	Browns
Observed, O	124	91	181
Expected, E	113.3	74.4	208.3
$O - E$	+10.7	+16.6	−27.3

The χ_2^2 is 8.29 with a $P < 0.02$; thus the deviations from expected may be judged significant. Cain and Currey choose to use the $O - E$ deviation of −27.3 for the browns as an estimate of selection in the following terms: ". . . selective differential of pinks and yellows combined relative to browns is about −27%." By using the cross-product ratio, the estimate of fitness for the browns is:

$$\frac{\text{expected pinks + yellows}}{\text{expected browns } (w)} = \frac{\text{observed pinks + yellows}}{\text{observed browns}},$$

and w is found to be 0.757, corresponding to an s value of 0.243. This value is the percent by which ideal fitness is reduced and relates to the value of 0.27 used above.

In these examples, the coefficient represents selection over an incomplete generation and the proportion of the generation involved may vary widely. A period of 24 or 36 hr no doubt represents a greater part of the *Biston betularia*'s adult life than of *Cepaea nemoralis*. Perhaps this difference has no meaning, since the interesting comparisons are between fitnesses of different morphs or between different populations of the same species. Nonetheless, data for one species may reflect from test to test different portions of adult life spans. Consequently, a standardized part of a generation is desirable for partial-generation coefficients.

Clarke and Sheppard (1966) appear to be the first workers to utilize this approach in selection studies. Working with the morphs of *Biston betularia*, they compared relative survival by way of the mean adult life expectancies, e_a, of the two major morphs. The rationale and assumptions necessary for obtaining life expectancy data from field estimates of death rates, d, were given in Chapter Two. The estimate was obtained by a summation process of successive samples; however, the interval between samples of natural populations cannot realistically be reduced below a one-day period. This limitation can cause the summation analysis to overestimate e_a if d is high for intervals between samples. The number of individuals alive at one time unit after observations begin is $1_1 = 1_0 s^1$ where s^1 represents a survival factor specific to time interval 1. If d is constant, s is also constant, and the number alive at time x is $1_x = 1_0 s^x$, thus $s^x = 1_x/1_0$. If d is rather large (Cook, 1971, suggests 0.2 or greater) and x intervals cannot be reduced, the life expectancy expression

$$\sum_{i=0}^{n=\infty} 1_x/1_0,$$

may overestimate the expectancy. These values also equal the area under the curve, $1_x = 1_0 s^x$, where 1_0 is the number at birth or number of emerging adults. The limits of the curve for either situation may be given as zero and infinity. Using integration to obtain this area, we have:

$$\int_0^\infty 1_0 s^x dx.$$

The 1_0 is a constant, so we obtain:

$$1_0 \int_0^\infty s^x dx \text{ or } 1_0(s^x/\log_e s + C),$$

and with numbers treated proportionately, 1_0 is 1.0, giving $s^x/\log_e s + C$. The term C is the integration constant. Subtracting the lower from upper limit gives $(s^\infty/\log_e s + C) - (s^0/\log_e s + C)$. When x approaches infinity, the $s^x/\log_e s$ term approaches zero, and we obtain $-s^0/\log_e s$ or $-1/\log_e s$ as the estimate of e_a.

Clarke and Sheppard (1966) applied these methods to the *Biston betularia* population around Liverpool, England. The basic part of their data relating to selection consisted of d_1 and d_2 of 0.80 and 0.52 for the typical and *carbonaria* morphs, respectively. These d values were obtained as described below in Bishop's (1972) study. The relative fitness of typicals to *carbonaria* is thus d_2/d_1 or 0.65. However, the e_a was judged to be longer than the actual period of reproductive activity by about one day. This one-day period did not contribute directly to the next generation and was subtracted from the e_a of each morph. The relative fitness of typicals then becomes 0.27. The loss of one day from a short-lived species has large impact on its reproductive contributions. Since a daily death rate of 80 percent is high relative to the sampling unit of time, i.e. a day, the authors suggest that the e_a of 1.25 days is an overestimate. Thus, using the integration estimate for e_a, $-1/\log_e s$, one obtains $-1/\log_e .2 = e_a$ of typicals as 0.62, and $-1/\log_e .48 = e_a$ of *carbonaria* as 1.36. The relative fitness of typicals is now 0.456. The restriction of reducing these expectancies by one day requires estimating the area of the life-expectancy curve between the limits of 1.0 and infinity. The expression is $-s/\log_e s$, and the expectations are 0.124 and 0.654, respectively. The fitness of typicals with these values falls to 0.189. The one day removed from expectancies may involve no reproductive activity; however, selection that day may well differentially eliminate individuals,and survival is the component of fitness that is being measured. Since m_x is taken as constant, this adjustment seems unnecessary. Thus the fitness estimate of 0.456 for typicals seems most acceptable and relates to a s of 0.544. This value is remarkably close to

Table 1. Comparison of survival numbers between *Carbonaria* and typicals

	Carbonaria			Typicals		
	Exposed	Taken	Surviving	Exposed	Taken	Surviving
Sefton Park	56	14	42	56	17	39
Eastham Ferry	80	20	60	80	28	52
Hawarden	52	22	30	52	18	34
Loggerheads	60	16	44	60	9	51
Llanbedr	60	23	37	60	16	44
Pwyllglas	84	40	44	84	20	64
Clegyr Mawr	92	39	53	92	24	68

estimates obtained in Birmingham by Kettlewell given above ($w = 0.434$ and $s = 0.566$).

Bishop (1972) continued Clarke and Sheppard's work with *Biston betularia* around Liverpool and obtained selection estimates along a cline leading to the outskirts of the urban area where polluted conditions were mostly absent. The d values were obtained by placing previously frozen moths on their natural resting sites, tree trunks, and observing the number missing after given intervals. The techniques were shown not to affect a bird's willingness to take such moths, and the trees were selected to represent random conditions. The daily death rate for each sample is (number taken)/(number exposed). For seven localities along the cline, numbers exposed, predated (taken), and surviving are shown in Table 1. The relative fitness of typicals to *carbonaria*, computed by cross-product and life-expectancy ratios, appear in Table 2. The former ratios use exposed and surviving numbers and the latter use the $-1/\log_e s$ modification.

Table 2. Relative fitness of *Typica* to *Carbonaria*

	Cross-product ratio	Life-expectancy ratio
Sefton Park	0.928	0.792
Eastham Ferry	0.867	0.666
Hawarden	1.133	1.295
Loggerheads	1.159	1.908
Llanbedr	1.189	1.559
Pwyllglas	1.454	2.376
Clegyr Mawr	1.283	1.823

The estimates given by Bishop were obtained by a different integration but agree well with these values numerically and in relative changes between localities. Actually, the fitness values taken directly from the cross-product method using observed numbers reveal equally well the trend in selection's magnitude. This study appears to be the only example where actual fitness values have been measured at various sites along a natural cline. Note that fitness does not seem to significantly change at any one point, as is assumed in the cline model discussed above.

SELECTION AGENTS

The impact of selection varies not only with the size of the s generated, but in temporal patterns of operation. In the discussion of non-neutral equilibria, two periods of selection were observed to be analyzed by summing the fitnesses, if different genotypes were affected each period. Namely, fitnesses of 1, 1, and $1-s$ and $1-s'$, 1, and 1 for genotypes AA, Aa and aa during two periods within a generation may be summed. On the other hand, $1-s$, 1, and 1, followed by $1+s'$, 1, and 1 within the same generation, most likely results in a net loss for the AA genotypes. The importance of such considerations at this point lies in how much selective loss is to be attributed to a specific agent of selection. An estimated selective loss per generation may not be a valid measure of one agent's effect. Take, for instance, two separate forms of selection developing s values of 0.2 and 0.3 in simultaneous operation against one genotype. If we take the A locus with $p_0 = q_0 = 0.5$, then the selection takes the following form:

Genotypes	AA	Aa	aa
Before-Selection Frequency	0.25	0.50	0.25
Fitness	$1-(0.2+0.3)$	1	1
After-Selection Frequency	0.125	0.50	0.25
Adjusted Frequency	0.143	0.571	0.286

The new value of p is found to be 0.429. If the two agents of selection act one after the other during the same generation, a different effect results:

Genotypes	AA	Aa	aa
Before-Selection Frequency	0.25	0.50	0.25
Fitness (1)	$1-0.2$	1	1
After-Selection Frequency	0.20	0.50	0.25
Adjusted Frequency	0.210	0.526	0.263
Fitness (2)	$1-0.3$	1	1
After-Selection Frequency	0.147	0.526	0.263
Adjusted Frequency	0.157	0.562	0.281

The new value of p is now found to be 0.428. In this numerical example, the overall difference of Δp is not great for the full generation; however, this difference could increase if the two s values change. The important point is that the effects are not equal if the agents act simultaneously or in series, and, if only one of the two agents were known, much of the Δp may erroneously be attributed to a nonresponsible agent. Environmental factors or factor interactions reducing viability or fertility vary from species to species and also between habitats. The likelihood of such complications as mentioned here is best judged by investigators who have a keen understanding of their subjects in nature.

The use of direct estimates for s values rests on the degree to which we can measure selection. One can predict the lowest s that may have effect on a population, recalling the necessary property of $s > 1/N$, and we may actually document the higher s values. From material in Chapters Two and Three and the present Chapter, the difficulty of statistically verifying small coefficients should be evident. The effect of genetic drift may usually be offset by lower s values than we can generally measure, so it is unlikely that we will distinguish between selection or drift by short-term studies. The s values we can measure will perhaps be of most use in identifying the mechanism balancing a polymorphism. High s values are generally associated with a balanced genetic polymorphism and perhaps can be used to distinguish between intrinsic heterosis and some ecologically imposed balance.

As selection acts on a population, it means fitness increases and the variance of phenotypic expression progressively decreases. The increase in fitness is greatest at high values of the variance. These relationships, explained in Chapter Seven, are important to recognize at this point as they determine the relative importance of different agents. When two or more selection agents operate in a sequential fashion on a character within a generation, the first agent acts on the greatest variance. It has therefore a higher likelihood of contribution to the increase of fitness than a second or third agent. The last agent to act must exert a relative higher s value to be effective.

Chapter 6

Varying Fitness and the Unit of Selection

Earlier chapters have discussed selection as a constant property, a view that is generally compatible with available data. Nonetheless, conceptual views of evolution argue that with data for longer periods or improved analysis, a varying fitness will be found. The role of varying s values in evolution has received attention because of their possible effects on two important issues in population biology. The first consists of balancing polymorphisms (non-neutral equilibria) without invoking superiority of fitness for the heterozygote, and the second concerns implications that such s values may have relative to conclusions suggested by genetic loads.

Theory does not require that all s values be vested with a varying property. If a recessive gene is lethal or causes sterility, etc., in the homozygous condition, then homozygotes formed by recombination will die, exist as sterile individuals, etc., regardless of habitat, population density, frequency of the gene, etc. This type of gene expression concerns what Wallace (1968, 1970) and Mather (1973) term *hard* or *unconditional selection* and appears to be equivalent to the notion that Turner and Williamson (1968) identify as loaded selection. Such selective loss is largely a matter of intrinsic physiological failure rather than of changing predator pressure, lack of sufficient refuges due to changing density, etc. A predator may well take such individuals, but they are already out of the picture for passing on genes anyway. In the discussion of non-neutral equilibria, attention was directed to the higher likelihood of balance by

137

way of intrinsic physiological effects on fitness rather than by environmental effects. These intrinsic reductions of fitness need not be total, i.e.
lethal, to still contribute to stable gene frequencies. A great deal of
selective loss is clearly not due to intrinsic failures. Non-hard selection is
therefore the concern here, and the changing s of most theoretical interest
is viewed as a function of phenotype frequency or population density, or
both, rather than as a change of selection agent through time or space.

The concept of frequency-dependent selection has received most attention, and Murray (1972) has recently presented a strong case for its
wide-scale operation. Basically, the rare type in a frequency distribution is
treated in a nonrandom fashion. The two possibilities most notably cited
involve predator-prey and mate choice phenomena. In the former, predators are assumed to take common-type prey in disproportionately high
numbers (apostatic selection), and in the latter case, the common-type are
chosen as mates in disproportionately low numbers. In addition to prey
and mate choice, Murray lists a number of other situations where fre-
quency-depent selection appears as a logical and, to some, a compelling
explanation. These possibilities include: self-sterility allele systems, sex
ratio, Batesian mimicry, disruptive selection, and disease-parasite-host interactions. The genetic effect consists of retaining two to several morphs in
the population as fitness changes reverse the trend for elimination of one
morph. As an example of its operation in symbols, consider the A locus
with dominance where two phenotypes exist in frequencies of $1 - q^2$ and
q^2. Let the fitness of each be $1 - sx_1$ and $1 - sx_2$, respectively, where s is
positive. Loss of fitness is controlled by the x value representing the
appropriate phenotype frequency. The relationships are as follows:

Genotypes	AA	Aa	aa
Before-Selection Frequency	p^2	$2pq$	q^2
		$1 - q^2$	
Fitness		$1 - s(1 - q^2)$	$1 - sq^2$.

Overall fitness of the phenotype increasing above its equilibrium point is
reduced, since the term consisting of the product of s and the phenotype
frequency increases; the reverse applies to the other phenotype. This effect
can be observed by considering the fitness of q^2 relative to that of $(1 - q^2)$, or $(1 - sq^2)/(1 - s[1 - q^2])$. From this expression, the fitnesses of
both phenotypes are equal, i.e., the ratio is 1.0 when $q^2 = 0.5$, and this
ratio becomes < 1.0 as q^2 increases but > 1.0 as q^2 decreases. The
equilibrium point for q^2 is 0.5. Cook (1971) discusses more general cases
of gene equilibria under frequency-dependent selection, but it should be

clear that the equilibrium point will depend on the number of alternate morphs. For instance, if three morphs exist, the equilibrium frequencies would be $0.3\overline{3}$. Thus, frequency-dependent selection alone allows a prediction of the equilibrium values from the number of alternate morphs. Self-sterility alleles and the sex ratio may be the closest approach.

The potential importance of frequency-dependent s values to evolution creates a definite need to document the operation in nature. The difficulty in measuring w's in natural populations with sufficient precision to reveal statistically verified changes has led investigators to adopt experimental designs. Population-cage data allow a number of controls and measurements not usually possible in a natural system, but distinguishing between a change in w based on gene frequency and one based on time is still difficult. Population cages of *Drosophila*, for example, can undergo different routes of selection where effort has been expended to present similar environments (O'Donald, 1971), and the cause remains obscure.

Harding, Allard, and Smeltzer (1966), working with lima beans, followed a process designed to overcome this difficulty. They initiated populations from inbred lines at different times and obtained contemporaneous cultures having few and many generations from a point of origin. The inbreeding tendency of lima beans causes a loss in heterozygotes, but they nonetheless stabilized at about 10 percent after five generations. The experimental arrangement was designed to distinguish between: 1) an increase of heterozygote fitness with time, and 2) a fitness dependent on frequency. Appropriate crosses between these stocks differing in age revealed no evidence for an increase of heterozygote fitness with time. The conclusion was that frequency determines fitness. As indicated above, an equilibrium may be established in an inbreeding system if selection acts against the homozygotes. The fitnesses are not changing in such a system. The frequencies of genotypes only change as they adjust to the relative fitness of heterozygotes and homozygotes. The lima bean study seems not to exclude this possibility, further indicating the difficulty of documenting frequency-dependent changes of fitness. Prout (1965) has also drawn attention to another possibility of misjudging the mode of selection. If frequencies before selection of the genotypes *AA, Aa*, and *aa* were 0.81, 0.18, and 0.01, respectively, and in the before-selection period of the following generation 0.64, 0.32, and 0.04, respectively, then *aa* appears to have experienced a frequency-dependent survival. However, fertility differences per individual without any frequency involvement can produce the same effect. Again, the importance of recognizing the before- and after-selection aspect of one's data and the extent to which the selection process is complete for a generation is seen.

Another approach involves predator-prey systems in which a predator group is offered "prey" morphs in different frequencies. The "prey" may be artificial, such as differently colored food particles, etc., and the design of data collection has been carefully developed to reduce various biases (Manly et al., 1972). Usually predators are thus seen to form search images of the more common type and then overlook or pass by the rare form. Clearly, this behavior generates frequency-dependent selection, and if the predator has the capacity for such activity, it is not stretching credibility to believe the process would also occur in nature. The impact such predation has on a prey in terms of s values depends on their ecological association. For instance, consider a prey with morphs A and B and a predator discriminating in such a way as to include A and B in amounts of 10 and 90 percent, respectively, of its total take. If the total number eaten is small in relation to prey available, the s value is also small. Consider that A and B occur in numbers of 1,000 and 10,000, respectively, and that the predator takes 0.1 and 0.01 percent of their numbers under two different ecological conditions. The results are tabulated below:

Morph	Prey, N	Percent Taken	Absolute Fitness	Relative Fitness of B
A	1,000	0.01	0.99	
B	10,000	0.09	0.91	0.9192
A	1,000	0.001	0.999	
B	10,000	0.009	0.991	0.9920

The case where predators remove only 0.01 of the prey generates an s value of only 0.008, or one 10 times smaller than reached when 0.1 of the prey are taken. This result illustrates a small selection effect due to a small s value, as discussed below. The possibility of other fitness variations swamping this effect is clear. Consequently, results of experimentally derived trends for selection must take into account the magnitude of ecological interaction when extrapolating the findings to natural situations.

In the case of mate choice, experimental confirmation is also positive, and is reviewed by Parsons (1967, 1973). In these test designs a choice is provided, and while test time and frequency of alternatives among available choices may be controlled, the approach to conditions in nature is speculative. To exercise choice, an individual must have instinctive preferences or have acquired them through some previous association (presumably the latter when choice is to be based on the frequency of neighboring types). An individual in the field, as it becomes receptive to mating, is most unlikely to have all choices available to it simultaneously. To exercise choice, the average individual may well have to by-pass mating and gamble on encountering the acceptable and receptive alternative later. The means of assembling sexes for mating in different species may reduce some of these difficulties in the exercise of choice, but it is difficult to accept the idea of such behavior in a wide variety of groups. Lewontin (1974) gives another view. If the level of heterozygosity in a population involves 30 percent of its loci and about 10 percent of loci in the average individuals, then the likelihood of genetically similar mates, say males, is low, and all males are rare. Students of frequency-dependent mate choice generally score by variation of only one character, and in the better-documented studies (Parsons, 1973 and earlier), the animals also discriminate in this simple fashion. Le Moli (1972) has recently recognized a property of *Drosophila* laboratory stocks which he describes as domestication. These laboratory populations respond differently from stocks more recently taken from nature and could inject a source of error if one extrapolates laboratory findings directly to a field situation.

The sex ratio was an early recognized case of frequency-dependent selection, and Fisher's (1930a, 1958a) explanation for the common tendency of a 1:1 value is approximately as follows. The basic observation is that each zygote is half male, half female; thus, the male and female contribution is equal in each zygote of the F_1 generation regardless of the existing parental sex ratio. Therefore, gene contribution by the sexes to the F_1 generation is relative to the sex ratio in the population of parents. When relative gene contributions from each sex balance, then:

$$\frac{\text{Number of genes given } F_1 \text{ by males}}{\text{Number of male parents}} = \frac{\text{Number of genes given } F_1 \text{ by females}}{\text{Number of female parents}}.$$

Since the number of genes given by males and females to the F_1 are equal regardless of the population sex ratio, the numerators in the above ratios are equal. If the population sex ratio is 1:1, then the male term less the female term equals zero. However, if this difference between terms is positive, then the majority of parents are females but proportionately more genes from males occur in the F_1. The reverse condition develops if the difference between terms is negative. Thus, where the above difference is positive, any tendency for a parent to have a preponderance of male offspring will achieve greater genetic representation in the F_1. Selection therefore moves the ratio toward a 1:1 value among parents since this ratio gives an equal representation in gametes at fertilization. The fitness of a given sex is thus higher when at low frequencies in the breeding population. This rationale probably explains the very common observation of sex ratios near 1:1 usually associated with systems giving 1:1 ratios in individual broods. Some species have sex ratios among individual broods not agreeing with this pattern. Terrestrial isopods, for example, frequently have several female types (see White, 1973). Two types characteristically have either a majority of males or females in their broods, while another type has broods of approximately 1:1 ratios. This complex of female types exists in populations having sex ratios often, but not invariably, 1:1, and in which intersexes frequently appear. Much is yet to be learned regarding selection on the sex ratios in such species.

Since adults of most species apparently strive to function as breeders, the actual adult population ratio is a close estimate of the ratio among breeders. If the adult population ratio is not 1:1, parents may still be equally represented in the F_1 gene pool. The most plausible explanation requires that the sex experiencing the higher, early mortality compensate by having its survivors engage in sufficient multiple matings for balancing the loss. Thus, a female may have 50 males and 50 females in her zygotes at fertilization. One male and five females may reach the breeding age, a ratio approximating the population adult ratio, assuming that all sibships experience similar mortality rates. If the average male could achieve five successful matings for each mating accomplished by the average female, then no advantage exists for parents that produce broods deviating from the 1:1 ratio at fertilization. If the percent survival for males is five times less than the percent survival for females, then each male should mate an average of five times more frequently than the typical female. A mating system of this kind can very likely result in several males mating with one female, a pattern that will of course not produce the necessary balance if the male simply mates five times more frequently than the female. Sperm competition and precedence will largely nullify the efforts of many such males. The capacity for sperm storage in the female and a female's possible reluctance to mate twice could reduce this problem, but these traits seem to be infrequently observed.

In a population where the sex ratio is 1:1, multiple male matings may also be required to achieve an average of one successful mating per male relative to females, owing to sperm precedence. If the ratio shifts so as to produce females in lower abundance, the prospects for balance seem less likely. A time problem intervenes. One female mating several males, each for a distinct brood, is possible only over a long time, barring very unlikely assumptions. The time consists of the interval for oviposition, gestation, or whatever, between the sequential matings. Pooling the matings complicates the likelihood of balance, as mentioned above. One female mating sequentially to several males is not the same as one male mating several females unless the female mates to one male only for each brood of offspring and does not store sperm. The requirements on the single female are restrictive, and they could also be complicated by a mortality rate that removed females before they effected the necessary balance of parental representation in the F_1. Where unequal sex ratios in adults occur, having more females than males is an easier system in which to balance the parental representation.

Since unequal adult ratios are coupled to an expected 1:1 ratio at fertilization, the adult ratio gives a prediction for differential mortality and mating activity. The former is more susceptible to analysis. Thus, an unequal sex ratio in adults (the breeders) is evidence of selection if we assume that the sex ratio of individual broods can have any genetic basis. Either selection is favoring parents that are contributing more of the F_1 sex currently in lowest frequency, and thus moving the ratio toward 1:1, or selection is differentially removing one sex during an earlier stage in the life cycle. The former mode of selection changes the ratio slowly (Bodmer and Edwards, 1960). If inequalities in the ratio are due to such changes, it seems unlikely that different populations of a species would be similar but unequal in the sex ratio. However, the latter observation can be expected if an intrinsic failure of one sex occurs. An argument exists for an unequal sex ratio at fertilization if one sex should require more parental expenditure of care, energy, etc. These "extraordinary sex ratios" are discussed by Hamilton (1967) and in a series of essays edited by Campbell (1972). The 1:1 sex ratios agree well with the condition predicted earlier for stability; namely, the equilibrium value of 0.5 exists for each type. The case of self-sterility alleles, also explained by frequency-dependent selection, is best known in certain plants for three-allele systems. Here, the alleles also reach equilibria at equal frequencies of $0.3\overline{3}$ (Mather, 1973).

Some organisms, e.g. many snails and earthworms, escape the sex ratio problem by adopting hermaphroditism, although generally avoiding self-fertilization. In such systems the ratio is always 1:1 and any two individuals are potential mates. A more complex system involves sequential hermaphroditism, where first one sex is functional, and then the same individual at a later age functions as the alternative sex. Warner (1975) explains these patterns on the basis of fecundity. Protandry (first functional as a male) could have evolved if female fecundity was highest with increasing age. Protogyny (first functional as a female) possibly evolved owing to low male fertility at earlier ages or low female fecundity at increasing age. The possibilities for balance of parental input in the F_1 are more complex in such a system but appear to require a 1:1 ratio at early and late ages. Frequency-dependent selection may well exist on the wide scale suggested; however, beyond sex ratio and self-sterility allele systems, additional effects may very likely be required to explain the specific equilibrium values of many stable systems.

Non-hard, or competitive selection may also operate by placing fitness as a function of population density. The term *density-dependent selection* is often used to describe this system; however, controversial interpretations of this usage are held by ecologists, so Wallace (1968, 1970) proposes the term *soft selection*. The disadvantage operates only when some resource is in short supply, and to involve selection one phenotype must be able to utilize resources more effectively than others. Thus, if two phenotypes, A and B, exist, and A is better qualified than B to utilize resource X, then: 1) no selection relative to X between A and B occurs if some resource restricts the density before a shortage of X develops; or 2) selection occurs if a shortage of X develops and B has the lower relative fitness. A continuation of the shortage results in the population becoming essentially all A. Most shortages of resources probably develop with high densities; however, scarcity of mates at low densities could involve competition for locating mates, in addition to any sexual selection. Where A and B coexist with a total density equaling the carrying capacity of the resource, and A is the rare morph, then its higher relative fitness continues as it becomes the common-type and until it has replaced B, a time when fitness relative to B has no meaning. In this case A now competes only with other A types because the population has become restructured in relation to the morphs and A bears all of the ecological load. Should the population be entirely composed of B, intra-B competition also imposes an ecological load if X becomes limiting. The latter two cases consist of competition, as most ecologists view the subject, i.e., without recognizing relative advantages between different phenotypes. Turner and Williamson (1968) discuss soft selection in terms of organisms contaminating the environment and one genotype (the optimum) experiencing less ill effect. In the above description the increasing contamination is equivalent to a diminishing resource supply. As an example, Wallace (1970) discusses competition for pupation sites in culture vials of *Drosophila*. The potential for soft selection in nature is extremely widespread. Andrewartha (1971) gives an engaging discussion of animal responses to resource shortage, and ecologists generally agree that shortages operate in some way to control population numbers. If populations exist at stable average densities through time, owing to an adjustment to finite resources, then selection operating on competitive ability should lead to monomorphic populations best suited to existing needs of competition. We do not find monomorphic populations and could conclude that selection does not operate on competitive ability or that other systems override the competition effect. Selection is not likely to be insensitive to density, and thus competition. The absence of monomorphic populations may also be explained by fitness values being determined by both density and genotype frequency. Mather (1969, 1973) gives the formal guidelines by which such selection would operate.

Consider the A locus with three genotypes, AA, Aa, and aa, each with a different competitive ability for a given density. The components of fitness due to this difference can be given as:

		Competitors		
		AA p^2	Aa $2pq$	aa q^2
Genotypes	AA	1	$1 + s_1$	$1 + s_3$
	Aa	$1 - s_1$	1	$1 + s_2$
	aa	$1 - s_3$	$1 - s_2$	1

In competition with itself genotype AA has unit fitness, but with Aa and aa it has $1 + s_1$ and $1 + s_3$, respectively. Likewise, Aa experiences a competitive relation to AA and aa of $1 - s_1$ and $1 + s_2$, respectively, again having unit fitness with like genotypes. Genotype aa then experiences $1 - s_3$ and $1 - s_2$ competitive relations with AA and Aa, respectively, and unity with other aa. The values of s are determined by a given population density; however, the mean fitness of a genotype depends on the frequency of its competitors. Thus, the mean fitness of AA is $p^2 + 2pq(1 + s_1) + q^2(1 + s_3)$, which reduces to $1 + 2pqs_1 + q^2s_3$. Likewise, the mean fitness of Aa and aa can be expressed as $1 - p^2s_1 + q^2s_2$ and $1 - 2pqs_2 - p^2s_3$, respectively. We have, therefore, the following expression of selection:

Before-Selection Frequency	Fitness	After-Selection Frequency
p^2	$1 + 2pqs_1 + q^2s_3$	$p^2(1 + 2pqs_1 + q^2s_3)$
$2pq$	$1 - p^2s_1 + q^2s_2$	$2pq(1 - p^2s_1 + q^2s_2)$
q^2	$1 - 2pqs_2 - p^2s_3$	$q^2(1 - 2pqs_2 - p^2s_3)$

The algebraic sum of the after-selection frequencies is one. The new p frequency can therefore be expressed as follows:

$$p_1 = p^2(1 + 2pqs_1 + q^2s_3) + pq(1 - p^2s_1 + q^2s_2)$$
$$= p(1 + p^2qs_1 + q^3s_2 + pq^2s_3).$$

If neither p nor q equals zero and if the new p equals the initial p, then the sum of the latter three terms in the parentheses must equal zero. Rearrangement gives:

$$s_3 = -(ps_1/q + qs_2/p),$$

and any three s values satisfying the expression give an equilibrium. The equilibrium requires, however, that the density-determined s values and the p and q values remain in adjustment. If the values are sensitive to density or non-density components of fitness, the likelihood of equilibria seems low. Thus, if selection coefficients are defined in terms of density, impact of selection is still influenced by the genotype frequencies. On the other hand, if coefficients are defined only in terms of genotype frequencies, density could theoretically be immaterial. In actual practice, the latter case probably doesn't occur. Turner and Williamson (1968), reasoning along these lines, introduce the term *population-dependent selection* to recognize interaction of density and frequency on fitness, and they suggest it is the most important of non-hard selection. As an example, they cite a paper by Ford and Ford (1930) on the butterfly *Melitaea aurinia*. Some unknown mechanism controlling numbers of the butterflies relaxed, and their numbers greatly increased bringing varieties into frequencies not otherwise known. The lima bean example discussed above in relation to frequency-dependent selection was also found responsive to density effects. These concepts of non-hard selection largely explain conflicts developed from the idea of genetic load. Essentially, selective loss due to balanced polymorphisms may well have been overestimated. In other words, the early interpretations of genetic load were built on hard selection models.

The terms r and k selection, defined in Chapter Two, should be commented on at this point. As defined, populations reflecting r selection are those in which individuals with maximal reproductive output possess maximal fitness, and populations shaped by k selection consist of individuals with maximal competitive abilities having greatest fitness. A misinterpretation is possible. By describing r-selection in the above terms, some writings assume that k selection does not favor high reproductive ability. Taking this position requires an ability to explain how selection can fail to reward maximal reproductive output. This problem is discussed below in relation to group selection; however, we might observe that individuals in all populations most likely experience selection for maximal reproductive output but only some populations live in circumstances where significant but superimposed competitive selection exists. This latter condition constitutes a less difficult usage of the term k selection.

Frequency-dependent selection, in the pure sense, observes a control of s values by phenotypic proportions, and population-dependent, or competitive, selection operates through both density and phenotype frequencies. Selection values may change, however, independently of phenotype frequencies or population densities by way of habitat variation. Rather than a long-term environmental modification leading to a cline through time, the changes of interest here concern cyclic habitat phenomena. The likelihood of such cycles producing selection response may depend on the duration of a generation relative to the period of habitat cycling. The importance of such cycles in selection is aptly stated by Mather (1973) as: ". . . the genetical adjustment of a population to its circumstance can never come earlier than in the offspring of the parental population which suffered the rigours of selection imposed by those circumstances." The after-selection component of the generation experiencing selection will, in fact, have gene frequencies determined by selection, but only their offspring encounter similar selection at a similar age.

The relationships are shown graphically in Figure 1. Generation intervals shorter than habitat cycles appear unable to develop specific genetic adjustments to each wave of environmental change. A switch mechanism, geared to environmental events, may, however, evolve to direct development of one genotype into different seasonal expressions. Major cycles usually correlate with seasons of the 12-month year and species mentioned above, often termed ephemerals, have two or more generations a year. Mather (1973) suggests that such life cycles will develop homozygosity (or genetic invariance in his usage). Species with annual life cycles can be expected to show maximal adjustment to the 12-month cyclic changes, and long-lived forms, perennials, will have their adjustments selected for long-term average conditions. Figure 1 gives four years and relative activity of an environmental component plotted under two conditions. The cyclic condition is represented by line x, while a trendlike departure of conditions is shown by line y. Four species' life cycles are shown below the figure, and the length of each line is the relative duration of the generation. The change of conditions x and y are assumed to carry potential differences in selection values. Species A and B are ephemerals and should express little genetic variability in relation to cyclic conditions depicted by x; however, the y conditions constitute long-term trends relative to A and B, and if y departs from the mean of condition x, a move into directional selection, in which genetic variability is of value, develops. Species C is an annual for the hypothetical habitat, and although well adjusted to the mean of x, it will also experience directional selection should a trendlike change, like y, continue, thus departing from the mean of x. Species D is a

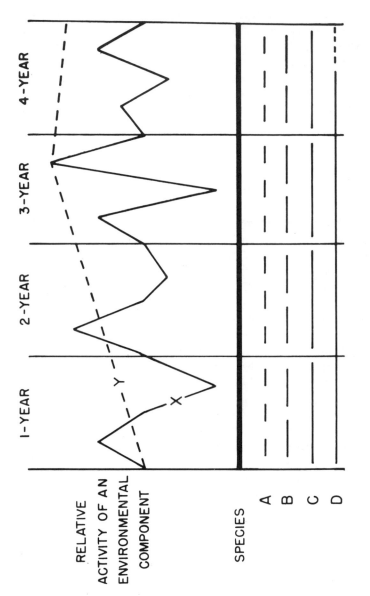

Figure 1. Generation intervals of four species, *A–D*, relative to a 4-year period. Species *A*, *B*, and *C* have 3, 2, and 1 generations per year, while species *D* is a perennial. Lines *x* and *y* denote an irregular and trendlike change in environmental condition. See text for interpretation.

perennial in relative terms and is not be expected to respond to either x or y conditions.

Recall that Haldane and Jayakar (1963) suggested the bivoltine (two generations per year) system with a varying s value as a possible means of balancing polymorphisms. An equilibrium developed in that way appears to entail considerable cost in terms of selection loss over the full year. Reproductive powers of many species seem capable of paying a high price in numbers, but any genetic variability not responsive to the changing s values and also leading to higher reproductive success per individual would be favored. Such variability seems likely to exist. Even so, some data for *Drosophila*, an ephemeral group, suggest response may be significant to seasonal changes (Dobzhansky, 1970 and earlier). The extent that natural bivoltine populations change genetically between generations is not sufficiently known for evaluation of these predictions. Perhaps polygenic systems respond to the full-year condition through stabilizing selection, while major genes or inversions segregating in unit fashion may be affected along lines suggested by Haldane and Jayakar.

The densities of populations frequently oscillate from high to low values through time. Directional selection could perhaps oscillate also where periods of low density serve as indices of periods having high selection coefficients. Haldane (1932) questioned whether intense selection operating periodically could produce a rate of evolutionary change different from that produced when selection was constant, but of less intensity, each generation. The result depends, of course, on the difference between the periodic and constant coefficient and the interval between the periodic effect. Haldane called such a system *cataclysmic selection* and found that the rate of change is greater if dominants are favored and is less otherwise, but the difference is small. If density oscillations were in fact caused by a system in which the coefficient varied only in intensity and period of operation, then the oscillation would theoretically diminish. The data of ecologists suggest that oscillations, while not highly predictable, continue to characterize most populations. Thus, times of low densities in an oscillating population are unlikely to reflect recurring high directional selection against the same phenotype.

UNIT OF SELECTION

The conventional and widely accepted concept of selection places the measure of success as the individual's contribution to the F_1's gene pool. The way selection acts upon individuals, by small, independent units or otherwise, is not an area of complete agreement. The unit of selection,

examined by an investigator, generally concerns a single character determined by one or two gene loci, an inversion, or a polygenic system. Fitness
itself is occasionally identified as the master character sensitive to selection. This master character is the composite expression of an individual's
total attributes and each specific character has a specific, optimal expression when total fitness is maximal. Specific characters may thus be used as
an index of selection; however, selective loss of each character may or may
not be independent. On this basis, selection may be observed by way of
any individual attribute. Also, the question remains: is observed variability
of any character without influence on fitness? Such characters, if they
eixst, would be neutral. The neutral concept is that the frequencies of
genes determining such characters are governed largely by genetic drift
(Crow and Kimura, 1970). Where data for several populations or sequential data on one population relative to a locus are available, neutrality is
usually rejected. The studies discussed above indicate the analyses possible
if selection is likely to be sufficiently intense for direct measurement or
for statistical comparisons of gene frequencies at the same locus in populations from different habitats. If genotypes are neutral, populations of
comparable size should have comparable genetic variances, but if they are
subject to selection, a process not likely to be similar in different populations, the variances should differ (Powell, 1971). Thus data may resolve
the neutrality question for specific traits, but the trait may not be a
random representation of a species' endowments. Consequently, one cannot argue that variation of most other traits is controlled by selection.

Such a conclusion can be invoked if the likelihood of any genotype
being neutral is shown to be low on other grounds. Fisher (1930b), as
indicated earlier, took this position at a time when the molecular nature of
the gene and mutation were less well known. A gene locus may be
completely homozygous, giving one expression for one trait or, by pleiotrophy, one expression each for two or more traits. The locus may be
heterozygous with two or more alleles, where two or more alternate
expressions exist for each trait depending on dominance in the allelic
series. These conditions are conventional gene properties and the key word
is expression. Selection recognizes expression. Fisher reasoned that selection does recognize allelic differences, although the selection coefficients
may be small, and he showed the probability was very low that an allele's
expression would exactly coincide with the zero point on an s scale from
minus to positive values. Thus, this position does not mean that selection
can't recognize alleles but one gathers that Fisher considered such alleles
to be unimportant.

The question now asked is: does selection fail to recognize a gene more often than Fisher believed? If a locus exists with two structurally different alleles giving, however, one expression, then selection presumably scores as one allele and frequencies of the two isoalleles come under other controls. The question of such truly neutral alleles largely developed with the advances in understanding gene structure. The macromolecular nature of a gene offers many sites for change, nucleotide-base substitutions, and many such changes are thought to confer minor or unimportant changes of the genetic message. Combined with this realization is the application of electrophoresis, which allows inferences about genes by the protein products found in enzymes. Collectively, geneticists now know that intragenic variation may not be insignificant, and the laboratory means for scoring this variation is quite sensitive. Perhaps it is more so than for selection. The likelihood for the latter possibility has, however, not received much support.

Powell (1971), Tomaszewski, Schaeffer, Johnson (1973), McDonald and Ayala (1974) and others have tested this possibility directly with electrophoretic methods and concluded that genes scored by their protein products were not neutral to environmental variation. Whole structures have occasionally been identified as neutral, often for the purposes of taxonomy, and such traits quite likely owe their expressions to polygenic systems. These interpretations are usually subjective judgements; however, genetic correlation between polygenically controlled expressions very likely explain many such characters, thus removing their immunity to selection (Falconer, 1960). Students of natural populations are justified at present in assuming that traits, which are recognizable by any method except perhaps the most sensitive of biochemical tools, are also recognizable by selection agents.

The question of what selection recognizes in the individual concerns the degree of scrutiny the individual can afford in relation to its reproductive output. The possible production of many suboptimal F_1 types, if many heterozygous loci are segregating, was mentioned earlier. In view of the available estimates of heterozygosity (about 10 percent of an individual's loci), the fecundity of many or most species seems too low to pay an independent price for each suboptimal genotype and still have enough offspring left. Thus, workers on evolutionary theory advanced the multiple locus hypothesis as an alternative to the single locus concept frequently implied as a unit of selection. The concept is well discussed in Lewontin (1974). Essentially, numerous loci are pictured as interacting and sorting into linkage groups that produce high frequencies of gametes, the non-crossover class, yielding the most fit zygote. Crossovers move linked loci toward linkage equilibrium; therefore, the loci interaction must be strong enough to override this force. In other words, linkage disequilibria must be permanently maintained. Since few gamete types are produced in relation to the potential number, judging from variation of individual loci, the diversity of zygotes is also reduced and the loss by segregation (segregational load) is much less than it is without the linkage effect. As selection operates to increase the tightness of linkage, the number of loci arrangements per chromosome pair reduces in number toward two. Thus, for a locus with two alleles the frequencies, p and q, approach 0.5.

This effect may be illustrated intuitively as follows. Many loci, arranged in linear order, occur along each homolog of a homologous chromosome pair. If allelic differences exist at a reasonable number of the loci, then the total number of different total chromosome arrangements is large. Even so, the system can be discussed as consisting of only three forms for each homolog, A, B, and C. Furthermore, the loci can be sorted into different sets owing to different arrangements of dominant and recessive alleles between different loci, or by way of multiple alleles at each locus. The diploid arrangements of the three homolog versions may exist in six combinations, A/A, A/B, A/C, B/B, B/C, and C/C. If one of the six combinations has an optimal, thus highest, fitness, say A/B, and linkage is complete, i.e., no recombination occurs, then the population will come in time to consist predominantly of A/B types relative to this chromosome pair. For each locus with allelic variation only two alleles remain, one in the A arrangement, and the other in the B arrangement. Since the population becomes restructured toward a point where most individuals are A/B, these two arrangements are about equally frequent and for any locus, $p \approx q \approx 0.5$. If the three chromosome forms involve multiple allelic variation, then recognition of only one combination like A/B as optimal results, as mentioned, in the loss of one allele per locus, that one carried on the C form of the chromosome in the example given. If all heterozygous combinations, A/B, A/C, and B/C, share the optimal fitness, then selection drives the system toward $p \approx q \approx r \approx 0.3\overline{3}$. If the optimal fitness is held by only a portion of possible heterozygotes, such as A/B and A/C, then selection leads toward stable frequencies in which one allele exceeds the others in frequency; in our example, $p \approx 2q \approx 2r$.

Should one of the genotypes sharing the optimal fitness class be homozygous, selection would move the population toward a monomorphic state because the offspring of the homozygous individuals would involve lower frequencies segregating into suboptimal classes. Actual data do not have many cases of monomorphic chromosomes. Homozygosity appears to be a poor solution for maximal fitness, judging from several lines of evidence. A shared optimal fitness involving whole chromosomes, such as $A/B = A/C = B/C$ or $A/B = A/C$, is also unlikely. Each chromosome form, A, B, and C, consists of many loci, and it is consistent with available data to believe that many of the loci have allelic variation. In reality, the number of chromosome forms will far exceed three. The likelihood that allelic variation between the chromosome combination A/B produces the same fitness as that generated by different alleles characterizing A/C or B/C is surely very low. Variation at only one of the many loci producing a differential effect on fitness can unbalance the above equalities and lead to only one true optimal combination, which returns the prediction to $p \approx q \approx 0.5$. If $p \neq q$ but nonetheless the frequencies appear to be stable over consecutive generations, then the explanation with fewest assumptions envisions the locus as being largely independent of other loci in its responses to selection. Namely, for the locus concerned, recombination is occurring and linkage disequilibria is not, controlling the frequencies of its alleles. Thus, the system may be suitably described in terms of single-locus models. Of course single-locus properties may also apply when $p = q = 0.5$, but other criteria will be needed to distinguish these cases.

Much of the data used for multiple locus discussions concern protein variation, recognized by electrophoresis, in which scored units are thought to relate, one for one, to single loci coding for the synthesis of polypeptide chains. Loci of most other genetic work are inferred from ratios of segregating factors in the F_1 from pair matings. Actually, the number of polypeptide-coding units in such factors is rarely known. The "loci" of different methods can clearly be different in genetic content. This possibility does not, of course, invalidate selection evidence traced to a locus. Most genotypes determined by alleles at a single locus probably contribute a small part of total fitness. As more loci are simultaneously considered, a larger fraction of total fitness is measured. Thus, significant differences in fitness are more likely to be observed in the simultaneous measurement of effects of multiple gene loci or of whole chromosome arms. This observation is not often helpful, because the more easily scored, thus more easily studied, traits are usually inherited in a single-locus fashion. The small input to overall fitness by many single loci make a statistical verification nearly impossible, owing to the practicalities of sample size. The exclusion of their input to total fitness becomes dependent upon the interpretation of the statistics.

Williams (1975) has questioned the need for a multiple-locus theory, at least in many organisms. He believes that many species, elm trees, various marine fish, etc., have sufficiently high fecundity for heavy selective loss without compromising an individual's representation in the F_1 gene pool. He also notes that if characters interact so that the fitness of one trait is dependent on another, a threshold effect results. This effect predicts no fitness increase when one character exceeds the threshold of maximum fitness. He contends that such a process is frequently not the case and cites performance data, particularly the high variability seen in fertility of a species, to support his position. He also observes that variation in fertility generally takes the form of asymmetrical distributions, and he gives statistical reasons for believing that these distributions could not describe a case in which fitness by fertility was largely dependent on other attributes. The actual mode of selection on the individual is thus a controversial issue.

If a multiple-locus situation exists, a practical question emerges: what parameters are actually reflected in the study of relative fitness differences between genotypes at only one locus? The nature of this problem may be illustrated using only two loci. If the A and B loci are considered, each with two alleles, then the fitness of each genotype per locus may be given as w_{AA}, etc. Likewise, the interaction of the two loci gives a fitness for each composite genotype, such as w_{AABB}, etc. These interactions are shown in the following way:

	AA	Aa	aa	
BB	w_{AABB}	w_{AaBB}	w_{aaBB}	$\overline{w}_{BB}$
Bb	w_{AABb}	w_{AaBb}	w_{aaBb}	$\overline{w}_{Bb}$
bb	w_{AAbb}	w_{Aabb}	w_{aabb}	$\overline{w}_{bb}$
	$\overline{w}_{AA}$	$\overline{w}_{Aa}$	$\overline{w}_{aa}$	W_T

The true fitness of each genotype per locus is w_{AA}, etc., but what is actually measured as individuals are scored, is $\overline{w}_{AA}$, etc., not isolated genotypes. If selection affects the loci independently, then w_{AABB} is equal to $(w_{AA})(w_{BB})$, etc., for the other combinations, and fitness is described as multiplicative, the $\overline{w}_{AA}$ equals w_{AA}. If w_{AABB} results from some epistatic interaction not multiplicative in operation, relative $\overline{w}_{AA}$ may not equal w_{AA}, but the observed values from studying individuals will equal $\overline{w}_{AA}$, etc., for fitnesses at the A locus. These mean values assume, of course, that the sample of genotypes for the A locus were proportionately distributed over the genotypes of the B locus. This assumption is true for multiplicative and other fitness interactions. Thus, when we measure fitness differences between genotypes at a locus, we are obtaining the mean values for each genotype. These values may directly reflect the independent fitnesses of the genotypes of the locus in question or they may depend on the frequencies of genotypes at other loci with which they do not have a multiplicative fitness interaction. In either case, the mean values we estimate for genotypes at a locus represent that locus' way of contributing toward the total fitness, W_T. The mean relative fitness between genotypes at a locus expresses their relative contribution toward total fitness regardless of the role, large or small, that the locus plays in total fitness. Thus, scoring individuals by a single locus does not mean that the locus is observed in isolation. The $\bar{w}$ for all genotypes at a single locus contributes differently toward W_T because fitness interactions of the different loci differ. Thus, the use of $\bar{w}$ values per locus should bear this relation in mind.

The possibility of a selection unit involving more than an individual's attributes has been raised several times (Sturtevant, 1938; Wynne-Edwards, 1962). The impetus for this position stems from observations that seem to be of: 1) altruistic or self-sacrificing behavior allowing survival or reproduction of others, and 2) an apparent or supposed advantage obtained by lowering one's reproductive output. The need for a mate also influences fitness. The typical reproductive pattern requires an individual to share half of its offspring's genes with another individual, its mate. In this way, a mated pair may be viewed as a unit of selection if loss of one member lowers fitness of the other (Mather, 1973). A few birds are thought to mate for life, and the formation of a second pair-bond by such an individual having lost a mate is questionable; however, most birds and other animals are willing to remate. Reduction of fitness in this way is not total but could involve loss of a full breeding season, and if such an individual dies before the next season its fitness would have fallen the maximal amount. Animal pairs with pair-bond behavior may thus represent a unit of selection, since the fitness of both individuals is affected by one act of selection. Actually, this property is a special case of the system in which pair-bonds occur. Sexual organisms require a mate, and survivors of pairs not likely to remate probably come to our attention more easily than unmated individuals in other mating systems. We rarely know what proportion of individuals fail to mate or how much variability may occur in that effort.

The reduction or increase of an individual's relative fitness consists of two components. These factors are: 1) the individual's attributes, and 2) effects from conspecific neighbors, or:

$$w = 1 - s_i - s_n,$$

where s_i is the loss due to individual attributes and s_n is the effect on fitness due to conspecific neighbors. The s_i values have a positive sign, or, if optimal conditions exist, equal zero. An s_n of zero describes a situation in which no effect on fitness can be traced to neighbors. The s_n with a positive sign reflects a neighbor-caused loss of fitness and is equivalent to the loss of a mate mentioned above or a competitive situation such as soft selection. Presumably the latter situation could develop at high densities, where a resource such as food or space becomes limiting, or also at low densities, where mates are in scarce supply. If the s_n value develops a net minus sign, the individual is then receiving favors, assets, or services from neighbors. Thus one individual is improving fitness of one or several other individuals. On superficial examination, this service frequently seems to lower the server's individual fitness. Hamilton (1964a, b) has shown that this impression depends on the degree of genetic relationship existing between the neighbors. For an individual the consequence of high fitness is the disproportionately higher frequency of its genes in succeeding generations than is seen for types with lower relative fitness. A gene's expression that favors its survival will, consequently, increase that gene's frequency. The expression may increase fitness of an individual, judged by the subsequent frequencies of genes it possesses, and yet place the individual in peril. The explanation lies in the measure of inbreeding. Briefly, if a locus of a parent is represented by genes A_1 and A_2, and two offspring both receive A_1 from this parent, the genes are termed, using inbreeding terminology, as replica genes or genes identical-by-descent, ibd. The relative proportion of genes ibd among relatives is Wright's Coefficient of Relationship and is 1/2 for parent-offspring and between full sibs, 1/4 between half-sibs, 1/8 between cousins, and zero where the relationship is negligible. Hamilton shows that much altruistic behavior is in fact an expression enhancing survival of the causative gene through its replicas in relatives. He expresses fitness in terms much like s_i and s_n, used above, defines it as inclusive fitness to recognize the neighbor effect, and develops a mathematical proof showing that the mean inclusive fitness increases with succeeding generations of selection, as Fisher's Fundamental Theorem shows for individual fitness (discussed below). Genes perpetuated in this way are said to reflect kin selection, and several interesting predictions result.

The advantage gained by sibs should theoretically equal at least twice the disadvantage incurred by the altruistic individual, the advantage for half-sibs should be four times the disadvantage, for a cousin, eight times, etc. Thus, no altruistic individual can increase its gene's representation by sacrificing its life for a single other individual, and distant relatives appear unlikely to receive sufficient advantage for encouragement of selection of altruistic expressions much beyond the immediate family. Traits giving personal gain that drastically reduce fitness of close relatives by taking their resources will not evolve. (These traits are "selfish" in Hamilton's usage.) At least half of the commodity taken from a sib must be used for reproductive advantage just to balance the loss inflicted on the gene's representation. Any increase in the gene's frequency can occur only by putting part of the second half into reproductive use also. Three-fourths of the taken commodity is available for use toward increasing the gene's frequency if the resource is taken from half-sibs. Net reproductive advantage for a gene is realized more easily if commodities are taken from distant kin so that the process becomes soft selection. Hamilton's papers suggest several important possibilities for evolution; however, two areas seem sufficient here for illustration of the complexities of kin selection.

Recall the difficulty for Batesian Mimicry in reconciling model mortality by predators to an individual unit of selection devoid of neighbor effects, and also, the altruistic element in Blest's suggestion of cryptic moths shortening their life expectancies so as not to compromise the next generation with educated predators. Hamilton suggests these cases, in part, as examples of kin selection. The unknown component for such examples consists of the coefficient of relationship with models and cryptic prey taken by predators. If the models and prey scatter randomly over a large area, intuitively the average relationship between those dying and surviving seems low. However, the possibility exists, within a conventional genetic framework, that selection favors a model's death or the "premature" death of cryptic parents. In socially oriented animals an obvious advantage develops for ability to discriminate, i.e., recognize simultaneously existing relatives. Evidence on this point is not lacking, and Hamilton also reviews these data. Still other cases exist in which kin selection is perhaps operating and altruism is not evident. In the African bombycid moth *Triloqua obliquissima*, the caterpillars crowd together on a leaf so that the effect is that they resemble a patch of excrement. The flattid planthoppers of the genus *Ityraea*, in this case adults, crowd together to resemble a flower. In one species, *I. nigrocincta*, two genetic color morphs exist, and the green individuals take positions near the top, the yellow individuals taking lower positions in the "flower inflorescence" (Fogden and Fogden, 1974). Species with warning coloration are on occasion seen to aggregate, presumably so that the group effect enhances the message to would-be predators. These examples clearly seem to associate a neighbor effect with fitness, so that the advantage develops from group participation only.

Colonial insects, particularly the hymenoptera, represent the extreme development of kin selection. Totally nonreproductive individuals, genetically females, are seen to forfeit their ability to have their own offspring and commit their lives to the rearing of their sisters. The reproductive system in terms of genes ibd make the system understandable. In hymenoptera, males are usually haploid and females are diploid.

For a cross such as $A^1 A^2 \times A^3$, the daughters are either $A^1 A^3$ or $A^2 A^3$. Genes ibd between mother and daughter give a relationship coefficient of 1/2; however, two sisters average three of every four genes ibd or a coefficient of 3/4. Thus, the female's likelihood of perpetuating genes of her type is made higher by raising a sister than it would be by raising a daughter. Presumably, a switch mechanism channels development of similar female genotypes into either reproductives or nonreproductives. The evolutionary pathway to such a mating system is discussed by Hamilton (1964b) and others.

Mated pairs thus involve positive or negative s_n values and the fitness of each member can be understood in a conventional framework of selection. Individuals engaging in kin selection presumably not only receive but also give in their own turn, and the neighbor effect is the net balance. For many species, the giving and receiving of favors may be separated according to age so as to produce a net minus s_n. A situation could exist in which a neighbor may act to increase the absolute fitness of unrelated individuals or lower its own absolute fitness. The effect is a decline in the relative fitness of the giver. The situation seems to have no explanation within our existing concept of selection. For instance, if some individuals adopt a course of activity that consists of producing fewer offspring than existing conditions allow, they experience a higher positive s_i than do individuals with more fecund behavior. If selection could favor the lower values within variability of reproductive output, population numbers could be regulated in relation to resource supply. The possibility has been discussed in the context of interpopulation or group selection. Wynne-Edwards (1962) gives the principle support for it. Maynard Smith (1964) and Cook (1971) discuss a possible origin for group selection. If a population becomes subdivided into small isolates where drift dominates the change of gene frequencies, then alleles conferring reduced reproductive effort, if available, could possibly be fixed. Two large problems exist, however. When the population begins to expand, the appearance of increased genetic variability is expected. If the variability encroaches on the loci having fitness-reducing properties, selection takes effect and the fitness-reducing genes are lost. Similarly, the fixation could not be expected to occur simultaneously in all habitats, and the influx by migration of genes having fitness-increasing properties would likewise lead to a loss of the population-controlling feature. These restrictions all but say that group selection cannot exist.

Against this strong case Wynne-Edwards (1963, 1964) offers the following observation: reduction in reproductive activity, or fitness, does not occur for all individuals. Some individuals are more easily deprived socially than others and for some reason do not mate. They seem to accept their reduced status and such behavior reappears each generation. The various behavioral and physiological components behind such a development must have some genetic basis. He asks how this attribute, judged non-fit, has continued to exist in a species when the concept of conventional selection predicts its elimination. A clear answer seems unavailable because the events under discussion are complex. Passive or submissive behavior is often an appeasement, and Lorenz (1963) outlines situations where it is of selective advantage. Even so, selectability of attributes for successful mating clearly exists, as shown by the known responses to sexual selection. If social dominance by only a portion of individuals is advantageous to local kin in areas in addition to that of merely obtaining a temporary mate, then kin selection may explain a number of observations. A switch mechanism, perhaps environmental, would channel development of one genotype into either submissives or dominants.

Another explanation for data that may appear to express group selection may lie in Goodman's (1974) comments and calculations on a "cost ceiling" in relation to reproductive effort. Goodman observes that reproductive activity often entails a risk and that maximal reproductive success may develop without maximum exposure to the risk. If only actual exposure relative to the possible maximal exposure is recorded, the data may resemble group selection. If the optimal exposure to risk is inversely proportional to conspecific density, a likely case in some groups (Johnson, 1964), then the effect would be a regulating influence on density.

Van Valen (1971) offers another view of group selection in terms of dispersal ability relative to the likelihood of a group going to extinction. Intuitively, the chance of an individual leaving a habitat in which conditions for existence and reproduction are good and finding another equally suitable place is low, quite low for certain species. Presumably, most dispersers are lost. The chance of survival and reproductive success for an individual disperser thus seems less than the comparable likelihood for a non-disperser. Habitats change and dispersal is necessary if extinction of a population is to be avoided. The proportion of any randomly chosen number of generations that experience a time when dispersal is mandatory is most likely a low value. So, selection would seem to favor parents whose offspring did not disperse. Nonetheless, dispersal occurs, and if the above assumption is valid, any genetic basis for dispersing would convey a survival value to the group, rather than the individual producing offspring with such tendencies each generation.

Boorman and Levitt (1972) and Gadgil (1975) also develop models, although somewhat restrictive, that explain group selection in terms of extinction and carrying-capacity of a habitat.

The explanation of sex in a diploid system has generally focused on the advantages of recombination in coping with the need for evolutionary flexibility as environments change. Maynard Smith (1971) has summarized several viewpoints, and he concludes that we do not yet understand sex, since some form of selection, approaching group selection, is suggested. The basic observations are as follows. A female producing diploid par-thenogenetic young leaves twice as many genes for a given number of F_1 as are left by a female sharing her offsprings' genes with a mate. The short-term advantage seems to be with the parthenogenetic female. Par-thenogenesis occurs in different groups and in different environments but not in frequencies expected by short-term selection effects; so a long-term effect of sex may not be included in the reasoning.

Williams (1975) describes the loss of gene representation in the F_1 gene pool due to sex as the cost of meiosis. He also believes that there are advantages to the individual due to sex that repay the meiotic cost, i.e., no long-term selection effect is necessary. He gives three basic models. He calls the first system the aphid-rotifer model, after the animals that may be involved. In such a system, habitats are discontinuous and discrete, individuals are motile and capable of dispersal as sexual diploids or asexual forms, and periodically a period of production occurs for dispersable units. Any genes that favor perpetuation of asexual reproduction may be called "asexual genes," and those allowing sex as "sexual genes." Dispers-able units produced by sexual reproduction are diversified because of segregation and recombination, while the units produced asexually, though numerous, are of one type. If an assemblage of units disperse into a habitat and begin to reproduce asexually, numbers will increase and competition will ensue. The initial period of reproduction favors an asexual ability owing to its rapid mode of multiplication, but as competition (soft selection) comes into effect, the likelihood is greater that one of the initial dispersive units, having an origin in sexual reproduction, will be the winner. Because environments are not constant, an unfavorable period can be expected, and the clone will again produce dispersive units. Since the winner of clonal competition during asexual growth is judged to originate from sexual parents, then it, too, will produce its F_1 by sexual reproduc-tion at the dispersing period. A balance is therefore suggested between asexual and sexual genes, both favored under different conditions. The advantages of sexual F_1 due to competition and environmental uncer-tainty are thought to outweigh the meiotic cost.

A second system, the strawberry-coral model, involves continuous or discontinuous habitats but sessile organisms with vegetative asexual growth, and the clonal life may be long, but not infinite. A clone's area is occupied by the best genotype suited to the area that has attempted colonization. It is limited at the boundaries by better-adapted clones or physical barriers. The area and biomass, or numbers, of the clone are determined by resources. Any resource channeled into sexual reproduction is a loss to the clone's asexual growth. Since conditions change, the existing clonal genotype is unlikely to meet all requirements. The loss of resources for personal growth plus the cost of meiosis must still be judged less than the cost of extinction. Thus, a balance again seems possible between sexual and asexual attributes.

Williams' third system is the elm-oyster model. Although many zygotes reproduce by asexual means and compete for a winning genotype of a clone in the earlier models, here many immature young compete for the ecological space held by one adult. Again, natural selection can determine the genotype of this winner. The period of competition is now between many sexually produced young and not between successive generations of asexually produced genotypes. The need for the "asexual genes" has vanished and such reproduction is usually absent. Williams suggests that once asexual reproduction is lost in such a fashion, it is largely impossible to reacquire. Therefore, systems in which asexual reproduction is absent but would seem more effective may be due to such an evolutionary past. Note that both Van Valen and Williams involve dispersal and chance of extinction in their reasoning. Van Valen suggests that dispersal's value is the prevention of extinction and that it is retained in a species by group selection. Williams accepts dispersal ability or tendency without questioning its relation to fitness, and he compares the prospects of sexually and asexually produced dispersers.

The models discussed above for explaining the value of sexual reproduction have several difficulties. All versions appear to rely on unstable environments, giving advantage to genes controlling sexual reproduction. If the environment remains sufficiently stable for a number of generations, the time when the sex genes would be advantageous is too infrequent and the genes could well be lost. Also, the inability to switch from sexual to asexual reproduction is not supported by the facts. Parthenogenesis, though not common, exists in several different groups, and its independent origin in each group is the most likely explanation. The difficulty is not an inability to assume asexual reproduction but the disadvantage of doing so. Much of the problem in understanding the cost of meiosis and the role of sexual reproduction versus parthenogenetic development may be in our way of viewing fitness. In Chapter Two, the possibility was mentioned that two morphs producing exactly equal numbers of offspring during their life times would not have equal fitness if the offspring of one morph were semi-sterile. Thus, two forms of fitness may be recognized. On the other hand, fitness may be judged by the relative number of offspring representing a morph's genes in the immediately following generation. This concept is usually followed, and it is for this that the algebraic symbolism for fitness discussions is constructed. If fitness were to be judged by the relative number of offspring representing a morph's genes two, three, or n generations in the future, then quite different relative values may result.

Parthenogenetic reproduction does increase the relative genetic representation of an individual practicing such reproduction over the value achieved by sexual reproduction, but the advantage may only apply to the very next generation. Parthenogenetic reproduction, in whichever of its various forms, involves inbreeding, and the depression of viability, fertility, etc., associated with inbreeding is well known for most organisms. For some species, inbreeding occurs regularly or periodically without a depression effect. Perhaps the cases of parthenogenetic reproduction involve species, e.g., the regular inbreeders, that have genetic architectures immune to the depression effect. Where inbreeding depression occurs, and data for its potential are at least numerous, the long-term view of fitness may well override the short-term advantage of parthenogenesis.

Today is the future for generations of the past. The individuals of two or more generations in the past that compromised their future representation for a short-term effect may well not be represented today, and their means of compromise may also be infrequent. Parthenogenesis is the usual departure from sexual reproduction and also the most likely pathway of possible compromise. While a long-term advantage for sexual reproduction in itself is difficult to formulate, a short-term disadvantage appears very likely for individuals that abandon sexual reproduction. This explanation seems more reasonable than a belief that asexual reproduction can't be developed from a sexual style even if the change confers higher short-term fitness.

Chapter 7

Quantitative Traits and the Selection Effect

Most investigators have pursued the study of selection using responses of qualitative, polymorphic variations. Such attributes are particularly suitable, since individuals can be rather easily sorted into distinct classes and are usually controlled by only one, or perhaps two, gene loci. Also, the expressions usually have rather little sensitivity to changing environmental conditions. Breeding tests or less direct population analyses can often identify the inheritance mechanism and thereby allow direct recognition of genotypes, which allows calculation of gene frequencies. Thus, specific fitnesses of single-locus genotypes may be obtained by way of one of several approaches given earlier, and Δq values provide direct measures of selection response. The data can also be expressed in the same units as usually appear in the theoretical models of selection. In addition, polymorphic variants often involve distinct expressions that readily suggest their value to individuals. In this way, the actual selection agent can frequently be identified. Thus, a specific genetic response in a population can be predicted if a known change in the make-up of the selection agent occurs.

Naturally occurring variation, particularly in the visible traits, is, however, frequently devoid of qualitative, polymorphic alternatives and consists of phenotypes varying continuously in a quantitative fashion. Actually, most visually recognized variation in natural populations is quantitative rather than qualitative, and a fact which surely plays a major role in determining the overall fitness of individuals. Thus methods for measuring selection on these attributes are required but entail different approaches. The selection agents acting on quantitative characters are usually more difficult to recognize, and the expressions are often sensitive to changing environments. In addition, the genetic basis for the expressions, discussed below, complicates the study because each unit on the quantitative scale may represent several different genotypes for the loci controlling the trait. Specific fitnesses for separate genotypes are generally unavailable. Selection loss must, therefore, be expressed as a mean value for an unknown number of controlling loci. Methods for measuring response to selection also differ for qualitative and quantitative variation. The response of qualitative characters was given generally in Δq terms while response for a quantitative phenotype is judged by changes in its mean and variance.

Directional selection produces significant Δq values or shifts a phenotype's mean. The difficulty in documenting significant Δq values, mentioned earlier, is due largely to small s values and small suboptimal frequencies at the time of study. Data for directional selection on quantitative traits no doubt will include the same difficulties. Also, any randomly chosen characters have most likely responded to directional selection and already have their mean values at the optimal point on the quantitative scale. However, just as a Δq of zero is not evidence of an absence of selection for qualitative traits, a non-changing mean value is not evidence of the absence of selection with quantitative characters. Namely, stabilizing selection may well be in operation, and the majority of data for quantitative traits do involve stabilizing selection. The various approaches to its study constitute the subject of this chapter.

A quantitative trait, such as size per age, bristles per segment, intensity of pigmentation, etc., appears to have its genetic determination vested in several gene loci or a polygenic system. Usually the number of loci in such a system remain unknown or can only be roughly estimated. The total phenotypic variation observed for such traits consists of genetic and environmental components. The genetic component can further be subdivided into additive and dominant gene action in relation to specific loci, plus an interaction component resulting from genes at different loci. If alleles A_0 and B_0 for two loci contribute to the expression of, say, small size, while alleles A_1 and B_1 contribute to expressing large size, then the full scale of genetically determined size relates to the range between $A_0A_0B_0B_0$ and $A_1A_1B_1B_1$. If the heterozygote, $A_1A_0B_1B_0$, develops an expression exactly intermediate between the two homozygotes, the gene action is said to be additive. On the other hand, if the expression of $A_1A_0B_1B_0$ is displaced toward one end of the range, a degree of dominance characterizes the gene action. The interaction component is more difficult to characterize; however, the available studies suggest that it forms a small part of the variation among polygenes and can, in a simplified discussion, be ignored (Falconer, 1960). Another interaction perhaps ignored with less justification is that of the genotype with the environment. These complications can, however, be dismissed for the purpose of examining the expected genetic response that selection should produce on polygenic systems.

Selection acting on a single gene locus leads either to: (1) fixation of one allele and consequent homozygosity, excepting the low frequency of variants due to mutational input, or (2) a balanced polymorphism resulting from heterozygote advantage, either intrinsically or otherwise. Selection acting on a quantitative trait is either directional, stabilizing, or disruptive in operation, as defined earlier. These selection patterns, while different in operation, have a similar effect, theoretically, on variation. Namely, they should lead to homozygosity at the loci determining the trait. The purest form of directional selection recognizes the extreme class at one end of the frequency distribution as optimal. If the variation is all additive, this optimal phenotype represents an optimal homozygous genotype, while a system where dominance exists may have several genotypes in the extreme but optimal phenotype. Thus, with additive gene action, pure directional selection favors a specific homozygote that will approach a state of fixation if selection continues. Where dominant gene action exists, segregation among the heterozygous dominants gives suboptimal types and selection still leads to homozygosity. Stabilizing selection in its ideal form recognizes the mode of the population's frequency distribution as optimal and reduces variability to either side of this value, i.e., it reduces the variance. If the optimal type is displaced from the population's modal value but is not the extreme class, then both directional and stabilizing selection exist simultaneously. Without invoking additional assumptions, this condition is unlikely to exist for long. The directional components of selection will move the population mode toward the optimal type and on coincidence only stabilizing selection remains. In the case of disruptive selection with, for simplicity, two optimal classes existing, each occurs as a point around which stabilizing selection operates. Thus, directional and stabilizing selection are the two basic processes. Directional selection, as indicated, leads toward homozygosity, but with stabilizing selection an intermediate class is favored that may involve several genotypes. Segregation among the F_1's of suboptimal parents may also result in optimal types. Thus, the move toward homozygosity under stabilizing selection is not as intuitively clear as it is under directional selection.

Formal proofs are discussed by Crow and Kimura (1970); however, the following simple example illustrates the expected process. If we consider two loci with alleles A_1, A_0 and B_1, B_0 that possess additive gene action, then we have a phenotypic scale of 0, 1, 2, 3, and 4 units, where A_1 and B_1 each contribute 1 unit per genotype. Thus, genotypes $A_1A_1B_0B_0$, $A_1A_0B_1B_0$ and $A_0A_0B_1B_1$ all have scale values of 2 each. The genotypes obtained from these two loci result from combining the four possible gametes as: $(A_1B_1 + A_1B_0 + A_0B_1 + A_0B_0)^2$. If the intermediate-scale class of 2 is taken as the optimum type, note that gametes A_1B_0 and A_0B_1 have two of four combinations, each expressing the optimal value, while the gametes A_1B_1 and A_0B_0 have only one of four combinations obtaining a value of two. Thus, the repulsion gametes contribute a greater part to the overall frequency of the optimal class.

If the frequencies of A_1, A_0, B_1, and B_0 are p, q, r, and s, respectively, the equilibrium gamete frequencies are pr, ps, qr, and qs. If p, q, r, and s are 0.5, 0.5, 0.6, and 0.4, respectively, and the population is initially at equilibrium for these loci, then the gamete frequencies, pr, ps, qr, and ps are 0.30, 0.20, 0.30, and 0.20, respectively. Inserting these values into the above expression gives the genotype values in column 3 of Table 1. Taking the scale class of 2 as optimal, we may, for simplicity, assign a fitness of zero to the suboptimal scale values of 0, 1, 3, and 4. The after-selection frequencies of the surviving optimals are then adjusted to a base of one and appear in column 4 of Table 1. We may calculate the gene frequencies after selection; however, recall from the discussion of linkage disequilibrium that gamete frequencies will not equal p_1r_1 etc. immediately following a disruption of equilibrium. This is true if the loci are not linked, a situation referred to as gametic equilibria. The gamete frequencies must be taken directly from the after-selection values. For instance, the frequency of A_1B_0 is 0.1081 + 0.6486/4 = 0.2703. These gamete values are given in the third row of the lower portion of Table 1 and provide a basis for obtaining before-selection genotype values in the following generation. The before-selection values are not shown, but again only class 2 survives, and their adjusted frequencies appear in the fifth column in the upper portion of the Table as after-selection frequencies for generation 2. In a similar fashion, the after-selection genotype frequencies for the surviving components of class 2 are shown for a third and fourth generation and the appropriate gamete frequencies for each of these generations appear in the lower portion of the table. Note that the move from equilibrium into the selection regime involves an increase in both homozygous genotypes for class 2 in the second generation. Beyond that time, however, the homozygous class determined by the repulsion gamete in what was initially the

Table 1. Values demonstrating development of homozygosity in stabilizing selection on a polygenic system with additive gene action[a]

Scale values	Genotype	Before-selection frequency Generation 1	After-selection frequency Generation			
			1	2	3	4
4	$A_1A_1B_1B_1$	0.0900				
3	$A_1A_1B_1B_0$	0.1200				
2	$A_1A_1B_0B_0$	0.0400	0.1081	0.1436	0.1339	0.1140
3	$A_1A_0B_1B_1$	0.1800				
2	$A_1A_0B_1B_0$	0.2400	0.6486	0.5337	0.5031	0.4824
1	$A_1A_0B_0B_0$	0.0800				
2	$A_0A_0B_1B_1$	0.0900	0.2432	0.3227	0.3629	0.4036
1	$A_0A_0B_1B_0$	0.1200				
0	$A_0A_0B_0B_0$	0.0400				

	Gametes Before-selection Generation 1	After-selection Generation			
		1	2	3	4
A_1B_1	0.30	0.1622	0.1334	0.1258	0.1206
A_1B_0	0.20	0.2703	0.2771	0.2597	0.2346
A_0B_1	0.30	0.4054	0.4561	0.4887	0.5242
A_0B_0	0.20	0.1622	0.1334	0.1258	0.1206

[a]Fitness values for scale units of 0, 1, 3, and 4 are taken as zero and the intermediate value of 2 has unit fitness. The population starts in equilibrium for the two loci where A_1, A_0, B_1, B_0 are represented by p, q, r, and s, respectively, having initial values of 0.5, 0.5, 0.6 and 0.4, respectively.

lowest frequency progressively declines in frequency, as does the double heterozygote. Likewise, beyond the second generation the only gamete frequency that increases is that of the repulsion gamete, initially in highest frequency. Since only one-fourth of the heterozygote frequency goes toward replacing each repulsion gamete, the repulsion gamete in the lowest initial frequency is destined to be lost. In the example discussed here, a homozygous population of $A_0A_0B_1B_1$ would be expected. This example uses additive gene action; however, dominance and incomplete dominance lead to the same conclusion unless the system consists of only a few loci in which inheritance is best described in other than polygenic terms (Crow and Kimura, 1970).

Pure directional selection as mentioned above probably occurs rarely. Such selection would expose a large frequency of suboptimal types to potential loss, and populations may not be able to withstand such extreme selection. Thus at any one time most selection on a quantitative trait is probably stabilizing. The fact that the frequency distributions of most continuously varying traits are near normal also supports this conclusion. An interesting comparison can be made between the actual degree that homozygosity is observed in such traits and the degree expected. Considerable evidence shows that the predicted development of homozygosity is rarely obtained. Fisher's (1918) paper first reported dominance in a polygenic system a condition revealing allelic variation. This conclusion was reached by noting that a higher correlation occurred between offspring of the same parents than between the parents and their offspring. Directional selection applied to quantitative traits under artificial conditions almost always achieves some response, again possible only if genetic variation exists at the determining loci. The studies on polypeptide variation by electrophoresis also suggest a large degree of variation, although the loci scored in that way may not characterize genes affecting many quantitative characters. We must conclude that the predictable trend to homozygosity must constitute a reduction in fitness and has apparently been opposed by selection.

If selection were to position genes into linkage groups so that the repulsion phase, A_1B_0/A_0B_1, was in highest frequency, the release of gametes A_1B_1 and A_0B_0, and thus not class 2 values, would be greatly reduced and the trend toward homozygosity impeded. If the genes were largely in the coupling phase, A_1B_1/A_0B_0, and if class 2 were still the optimal type, selection would take a heavy toll. If another class were optimal, the frequency distribution would be asymmetrical. This conclusion about distribution shape assumes additive gene action; however, if any existing dominance were equally partitioned to genes increasing and decreasing values about some intermediate value, then the conclusion about a symmetrically shaped distribution still applies. Where data exist, they show dominance for both high and low expressions of a quantitative trait (Mather and Jinks, 1971; Mather 1973). The common observation of symmetrical distributions for traits in natural populations and the fact that stabilizing selection, where studied, seems to be most common have played an important role in leading Mather to suggest balanced polygenic systems. If selection and linkage hold repulsive gene combinations together to give the gametes that consistently produce higher frequencies of the optimal type, then such systems offer a compromise between the needs of present fitness and future flexibility. This concept is discussed by some current workers as the multiple-locus concept of selection. Reference to future flexibility may appear inconsistent with the view of fitness in terms of present offspring survival and reproductive success. However, living species today are descendents of successful species or populations of the past. Where present species have been studied, the genetic architecture does approximate the balanced polygenic concept, and Mather argues that these systems explain the maintenance of variability in polygenic loci. Crow and Kimura (1970) and others observe that other views are possible. Small changes from an intermediate mean scale value may involve small fitness changes. Therefore, a modest mutation rate may introduce a significant degree of variation before the stabilizing selection balances its input. They also suggest that polygenes could have pleiotropic effects involving hetero-zygote superiority.

Disruptive selection is the eventual effect of directional and stabilizing components on homozygosity, but it also involves gene flow between the different optimal types. This flow can impede the development of homo-zygosity while its curtailment could theoretically fragment the complete gene pool. If the optimal types are separated in time, the system is not usually identified as disruptive selection since the types are not simulta-neously exposed to selection. The situations discussed above under cycles apply. Coexisting optimal types may be functionally dependent on each other, as in a species having two Batesian mimics. One morph mimics one model, a second morph mimics a second model, etc. If one morph becomes too numerous, the appropriate predators begin to confuse the image with edibility and unpleasantness. Thus, the success of one morph depends on the presence of the other unless the mimic's reproductive output can be selected to decline, an unlikely possibility.

Two optimal types related in this way need to have only minimal genetic differences, such as one or two gene loci controlling a particular color pattern or controlling alternate routes of later development, in order to retain themselves. Such control is known as a switch mechanism and the genes at other loci may not differ between the morphs. The evolution of switch genes is largely explained in this way (Mather 1973; Thoday, 1972). On the other hand, if the coexisting optimal types occur in different portions of the habitat, a dependent functional tie between them may not eixst. Genetic differences at many loci, at least differences in frequencies, may develop in association with such optimal types. The frequencies of the optimal classes will naturally increase within areas where selection favors each, and thus a bimodal fitness distribution gives rise to a bimodal phenotype distribution. This condition sets the stage for speculation on speciation closely approaching a sympatric condition, and the likelihood of this possibility is reviewed by Thoday (1972). Mayr (1974) objects to Thoday's use of disruptive selection where the optimal types are geographically separated; Thoday's (1974) reply stresses the fact that we cannot use objective definitions for the spatial area of most populations without becoming biologically inaccurate. This topic is beyond present objectives; however, recognition of the scope of disruptive selection brings otherwise diverse phenomena together. If selection operates differently in subsections of a general habitat in which random mating exists in each subsection, gene frequencies in each area will naturally differ. If a sample is taken from the full habitat area, the statistics produce a larger proportion of homozygotes than the mean gene frequencies predict for random mating. This phenomenon is known in population genetics as the Wahlund Effect, and was described earlier. Most references to this effect concentrate on the statistical parameter of genetic variance and its similarity to inbreeding; however, the biological question of most interest concerns the degree of gene exchange between the different and spatially separated optimal phenotypes.

Another related situation concerns Levene's model for habitat heterogeniety and a balanced polymorphism, mentioned above. Again, an integral part of this model is disruptive selection with the optimal types spatially separated. The model assumes individuals converge randomly to mate. A functional dependence between each optimum may be envisioned since the model assumes all mate together. Thus a common pool of alleles is shared by all, though at different frequencies. If mating is not fully random over the complete habitat, a more likely case than the system assumed by Levene, a departure from the common pool of alleles by one subdivision need not be lost or swamped. Murray (1972) suggests that frequency-dependent selection is in fact occurring in cases described as disruptive selection. Thoday and his colleagues selected for low and high lines in counts of bristle number from stocks of *Drosophila melanogaster*. These flies were housed in laboratory cages or culture vials and selection was imposed by the investigators. As selection proceeded, a bimodal response developed but one type usually outnumbered its opposite. The investigators selected equal numbers from both high and low lines as parents, introducing the complication of a frequency-dependent component. In a natural situation, where the two (or more) optima were spatially separated, each type would occur in highest frequency within its favored locale. This system would be unlikely to involve frequency-dependent selection unless gene exchange between such sites is large. The exchange of genetic material between spatially separated centers of optimal phenotypes appears to be the central question concerning the evolutionary importance of disruptive selection.

Selection estimates discussed in earlier chapters relate to distinct morphs determined by single gene pairs or similar segregating units. With quantitative traits such as body weight, height, length, etc., assigning fitnesses to separate genotypes is rarely possible. The analyses must pursue, therefore, estimates of mean selection values. The rationale employed can be illustrated with genotypes segregating from a single gene locus. When heterozygotes, for example, are assumed to have optimal fitness, we express selection as:

$$p^2(1 - s) + 2pq + q^2(1 - t) = 1 - (sp^2 + tq^2).$$

The fitnesses times genotype frequencies summed equal the mean relative fitness, $\overline{W}$, and we have:

$$1 - \overline{W} = sp^2 + tq^2.$$

Since $1 - w = s$, the selection coefficient, then $1 - \overline{W} = \overline{s}$, the mean selection coefficient of several loci, or, for our example, $\overline{s} = sp^2 + tq^2$ for the three genotypes of one locus. In systems with many loci segregating, many genotypes are formed, but we may assume one phenotype from an array of many possesses the maximal, thus optimal, fitness, w_0, although it may include several genotypes. The proportionate loss of fitness is $(w_0 - \overline{W})/w_0$, and where the convention of expressing w_0 as 1.0 is followed, the loss is again $1 - \overline{W}$, or $\overline{s}$. The s value for each suboptimal type most likely varies, and it increases for phenotypes that are progressively less like the optimum. A mean s is thus dependent on the relative frequencies of the optimal and suboptimal phenotypes and, indirectly, on the actual gene frequencies. A specific s value, as envisioned in earlier discussions, was determined by the selection agent. In frequency- or density-dependent selection, the s value varied because the selection agent responded to the frequency or density of the target morph. The $\overline{s}$ varies not because of the selection agent but because of the relative frequencies of optimal and suboptimal phenotypes. With the $\overline{s}$ concept, the overall loss of the population is estimated. This value is of course less than the loss of actual suboptimal types and declines as the frequency of suboptimal types declines. This distinction must be retained in interpreting the s and $\overline{s}$ values in evolutionary theory.

If the optimal phenotype of a quantitative trait is recognizable and before- and after-selection frequency distributions are available, the difference between the distributions can be analyzed for an estimate of $\overline{s}$. Weldon

(1901, 1904) and di Cesnola (1907) appear to be the first authors to report quantitative evidence for stabilizing selection. Basically, they found smaller standard deviations in adult samples than in juveniles for characters in snail shells. They could not, however, give estimates of the magnitude of selection involved. The goal of assigning values to such selection was developed approximately fifty years later by Haldane (1954). He used $S_0 - S$ as a measure of selection intensity, where S_0 is the percent survival of the optimal type and S is the percent survival of the whole population. Selection intensity, defined in this way, can be shown to be equivalent to the $\bar{s}$ as obtained above. If the proportionate loss of optimal types is α, then $1 - \alpha = S_0$. Likewise, $1 - \beta$ can be given as the surviving value of suboptimal types. Then if C and D represent the frequencies of the optimal and suboptimal types, respectively, where $C + D = 1.0$, then $S = (1 - \alpha)C + (1 - \beta)D$. Thus:

$$S_0 - S = (1 - \alpha) - [(1 - \alpha)C + (1 - \beta)D] = -D(\alpha - \beta).$$

If for simplicity we take a single gene locus where heterozygous advantage exists, then $\alpha = 0$ and $S_0 - S = D\beta$. The selective loss now equals $D\beta$, but this is the loss due to the two homozygous genotypes. With heterozygous advantage, we have:

$$p^2(1 - s) + 2pq + q^2(1 - t) = 1 - (p^2 s + q^2 t) = \bar{W}.$$

This loss is seen to be $p^2 s + q^2 t$; therefore:

$$S_0 - S = p^2 s + q^2 t, \text{ or } S_0 - S = 1 - \bar{W}, \text{ or } S_0 - S = \bar{s}.$$

Haldane in practice chose to use the $\log_e$ transformation as his parameter for the intensity of selection. Namely, his intensity estimates were:

$$\log_e S_0 - \log_e S, \text{ or } \log_e S_0 / S.$$

The $\bar{W}$ and $\bar{s}$ values can of course be computed directly from sample data if w and s values for each class are available. Such data are rarely obtained, but a hypothetical example is useful in illustrating the process. The operations are shown in the curves of Figure 1. Curves I and II_A are before- and after-selection frequency distributions, respectively. Maximal survival occurs at x'. The after-selection distribution is adjusted proportionately, giving x' a 100 percent survival, i.e., curve II_B. The difference between curves I and II_B, the shaded area, is the proportion lost by selection, or $\bar{s}$ when one assumes a fitness of 1.0 for the optimal type. In this case, the full range of x is assumed to be represented in the distributions. The illustrated curves approximate normal distributions in shape,

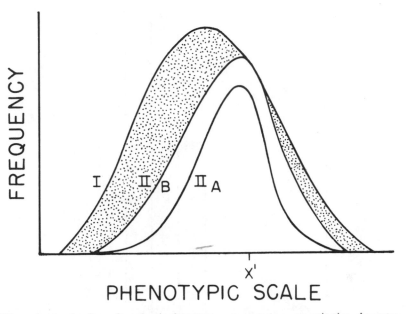

Figure 1. Application of optimal phenotype concept to a quantitative character. Curves I and II*A* are before- and after-selection curves identifying x_i as the class with optimal survival. Curve II*B* adjusts the after-selection distribution, giving x' survival of 1.0 and other classes in proportionate amounts. The shaded area between Curves I and II*B* is that part lost to selection.

but while these conditions do not affect the definition of $\bar{s}$, they are important in making estimates from sample data.

To illustrate the estimating process, Figure 2 gives a large and small histogram representing before- and after-selection frequencies, respectively, for a hypothetical organism having a trait varying from 13 to 22 scale units. The before- and after-selection samples are 950 and 515, respectively. The figures above each bar give actual frequencies. The percent survival (absolute fitness using observed numbers) appears in the first row below the corresponding class. The second row gives proportionate numbers when the class (scale value of 17) having maximal survival is adjusted to 100 percent survival. Relative fitnesses, computed by the cross-product method for each scale value, are shown in the third row. Variances of before- and after-selection distributions, 3.9668 and 3.2184, respectively, give an F-test value greater than 1.0 for 949 and 514 degrees of freedom, respectively, thus confirming a significant reduction of variability during selection. The after-selection frequency is adjusted by multiplying the frequency of each class by the factor necessary to give the

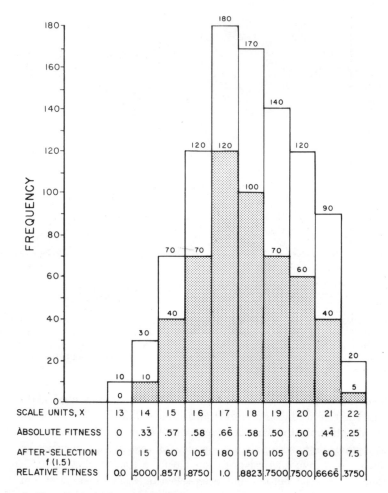

Figure 2. Hypothetical histogram illustrating computations of selection intensity, I. The upper, clear bars give before-selection, and the lower, shaded bars, give after-selection values, respectively. See text for explanation of figures.

optimal class the same frequency as it possessed before selection. In this case, the factor is 1.5, since $(120)(1.5) = 180$. The adjusted, total after-selection sample is 772.5, with 177.5 lost; or, in other words, 18.68 percent were lost in selection. A cross-product computation for $\bar{s}$ using totals and optimals also gives 18.7, or ($w950/515 = 180/120$). Table 2 illustrates computation of the same value using relative fitnesses of each class and class frequencies following the genotype-fitness notations in earlier chapters. In actual practice, relative fitness values for separate classes will not be available, and four expressions have been mathematically solved to overcome this problem.

Haldane (1954), using the rationale for selection intensity explained above, developed the following term for his estimate of selection intensity, I^H:

$$I^H = \log_e \frac{SD_B}{SD_A} - \frac{(\Delta \bar{X})^2}{2 \Delta V},$$

where SD_B and SD_A are standard deviations of the before- and after-selection distributions, $\Delta \bar{X}$ is change in means $(\bar{X}_A - \bar{X}_B)$, ΔV is change in variances $(V_A - V_B)$, and e is the base of natural logarithms. The sign preceeding the second term of I^H differs from Haldane's usage owing to the definition here of $\Delta \bar{X}$. If no significant difference exists between the means (ideal stabilizing selection), then I^H simplifies to $\log_e(SD_B/SD_A)$. Using Haldane's basic expression for selection intensity, $\log_e S_0/S$, if S_0 is taken as 1.0, as is conventional, I^H equals $-(\log_e S)$, since $\log_e S_0/S = \log_e S_0 - \log_e S$ and $\log_e S_0$ = zero. Of course I^H itself is positive because S is fractional, and therefore its logarithm is negative. This means that I^H does not have limits of zero and 1.0 as a conventional coefficient does. The antilog of Haldane's expression is e^{-I^H}, and since e^{-0} = 1.0 and

Table 2. Direct estimate of $\bar{W}$ and $\bar{s}$ by relative fitness[a]

Scale value x	Frequency f	Relative fitness w	f/N	(f/N) (w)
13	10	0	0.01053	0.00000
14	30	0.5000	0.03158	0.01579
15	70	0.8571	0.07368	0.06315
16	120	0.8750	0.12631	0.11052
17	180	1.0000	0.18947	0.18947
18	170	0.8823	0.17895	0.15789
19	140	0.7500	0.14737	0.11053
20	120	0.7500	0.12631	0.09472
21	90	0.6667	0.09474	0.06316
22	20	0.3750	0.02105	0.00789
	950 = N		1.00000	0.81312

$$1.0 - \Sigma(f/N)(w) = 1 - \bar{W} = 1 - 0.8131 = \bar{s} = 0.18688$$

[a]Example given in Figure 2.

approaches zero as the exponent increases, the expression $1.0 - e^{-I^H}$ has limits of 1.0. Thus Van Valen (1965) gives the expression:

$$I^V = 1.0 - e^{-I^H}$$

as a desirable estimate of selection intensity. Van Valen's estimate is equivalent to $(S_0 - S)/S_0$ and this value is a measure of the shaded area between curves I and II_B in Figure 1. In Haldane's usage curve I is the optimal performance, and if it is given a survival percentage of 1.0, it has a $\log_e$ value of zero; thus I^H is the negative logarithm of the area under curve II_B.

Van Valen's calculation of intensity relates to truncation (removal of extreme values to either end of the before-selection distribution) required to explain $\Delta \bar{X}$ (if any) and V_A. Both I^H and I^V also assume the distributions to be approximately normal. O'Donald (1968) suggests that Van Valen's truncation model of selection is unrealistic; rather, fitness as described above decreases progressively as a character departs from the optimum. He therefore modifies Van Valen's expression giving a third form for intensity, I^O, using the concept that fitness progressively decreases as the phenotype departs from the optimum (quadratic fitness in his usage). He also derives an expression of I^O for distributions departing from normality. His estimate would estimate $\bar{s}$ more accurately when both directional and stabilizing selection were coexisting. This estimate of intensity requires computing the third and fourth moments about the mean of the before-selection distribution, U_3 and U_4, in addition to terms required for I^H. The expression is more complex than I^H or I^V and is best given by identifying two terms in units of distribution statistics. If the optimum phenotype is X_0, the expression $X_0 - \bar{X}$ is:

$$\frac{V_B{}^2(\Delta \bar{X}) - U_4(\Delta \bar{X}) + U_3 \, [(\Delta \bar{X})^2 + \Delta V]}{2V_B \, [(\Delta \bar{X})^2 + \Delta V] - 2U_3(\Delta \bar{X})} \, .$$

X_0 does not depend on maximum observed survival identifying the optimal phenotype since this phenotype may not exist within the range of scorable categories. The fitness computation used by O'Donald involves a computed constant, ϕ, that can also be expressed in units from distribution statistics as:

$$\frac{2V_B(X_0 - \bar{X}) - U_3}{\Delta \bar{X}} + (X_0 - \bar{X}^2) + V_B \, .$$

From these terms:

$$I^O = \frac{(X_0 - \bar{X})^2 + V_B}{\phi} \, .$$

If the distributions are approximately normal and $\Delta \overline{X}$ is not significantly greater than zero, then:

$$I^o = \frac{V_B - V_A}{3 V_B - V_A}.$$

Cavalli-Sforza and Bodmer (1971) present still another intensity form, or:

$$I^{CSB} = 1.0 - \sqrt{V_A / V_B},$$

where $\Delta \overline{X}$ is not significantly greater than zero. For the hypothetical example above, the following estimates result. The direct measurement was 0.187 as given, and if we assume $\Delta \overline{X}$, (−0.101), is insignificant, then I^H and I^V equal 0.1049 and 0.0996, respectively. Incorporating the second term of Haldane's expression gives 0.111 and 0.105, respectively. The comparable I^o values are 0.0836 and 0.0917; the I^{CSB} estimate is 0.0996. The estimates show good agreement but are less than the direct measurement by an approximate factor of two. Table 3 gives a series of I estimates, and their general similarity to s values, discussed in preceding chapters, are apparent, excepting the cline-model values. These estimates of $\bar{s}$ appear to be higher than expected if the variability of polygenic systems were only maintained by a balance between mutation and selection, as one possibility mentioned above.

Quantitative traits as they are usually scored are phenotypic expressions with an unknown genetic component. O'Donald (1968, 1971) identifies I or $\bar{s}$, therefore, as the "phenotypic load," recognizing its similarity to the concept of genetic load. These estimates usually represent only a portion of the organism's life cycle and, for most data, give partial $\bar{s}$ values. These estimates may vary between sexes, suggesting that, where possible, data are best examined with this possibility in mind (O'Donald, 1971). The model for estimating I assumes one optimal class. If survival shows a bimodal fitness pattern or disruptive selection, the expressions for I are not valid.

Another approach to measuring selection consists of recording a population's response rather than the magnitude of s. Applied genetics has several such methods but only one concept has received much attention in evolutionary literature. Fisher (1930a, 1958a) drew attention to the rate of change in fitness and his observation develops in the following way.

When q is small, q^2 is also small and higher genotype frequencies develop for p^2 and $2pq$ (using the example of a single gene locus). Fitness may be envisioned as a variable for which the frequencies of its values are dependent on the genotype frequencies. Thus, the fitness distribution becomes progressively skewed as q decreases. Consequently, the variance

Table 3. Estimates of selection intensity, I or $\bar{s}$, and $\Delta\bar{W}/\bar{W}$ from natural populations

Species	Character	Intensity	$\Delta\bar{W}/\bar{W}$	Authority
Man	birth weight	$I^H = 0.024$		Haldane, 1954
	1st 28 days			
"Ducks"	egg weight	$I^H = 0.100 - 0.0276$		Haldane, 1954
Clausilia laminata, snail	whorls of shell	$I^H = 0.12$		Haldane, 1954
Thamnophis radix, snake	ventral scale no., males	$I^H = 0.136; I^V = 0.127$		Cook, 1971
Drosophila melanogaster	bristle no.	$I^O = 0.1078$	0.0252	O'Donald, 1970b
Maniola jurtina	spot no.	$I^O = 0.1275^a$	0.0423	O'Donald, 1970a
Capaea nemoralis	shell diameter	$I^O = 0.0381 - 0.136$	$0.003 - 0.027$	Cook and O'Donald, 1971
Cepaea nemoralis	shell diameter	$I^O = 0.225$	0.050	Bantock and Bayley, 1973
Cepaea hortensis	shell diameter	$I^O = 0.144$	0.015	
Gasterosteus aculeatus, threespine stickleback	lateral plates	$I^O = 0.469$	0.107	Hagen and Gilbertson, 1973
Passer domesticus, sparrow	gill rakers	$I^O = 0.452$	0.028	
	total length	$I^O = 0.195$	0.072	O'Donald, 1973
	humorus	$I^O = 0.185$	0.103	O'Donald, 1973

[a]The direct estimate is $I = 0.162$.

of fitness is less when q is small than when p and q are close in value. For example, if q is 0.1 and 0.5 in populations 1 and 2, respectively, and selection develops so that p^2, $2pq$, and q^2 have fitnesses of 1.0, 0.75, and 0.50, respectively, the following conditions exist:

	Population 1			Population 2		
Genotypes	p^2	$2pq$	q^2	p^2	$2pq$	p^2
Frequencies, f	0.81	0.18	0.01	0.25	0.50	0.25
Fitnesses, w	1.0	0.75	0.50	1.0	0.75	0.50
fw	0.81	0.135	0.005	0.25	0.375	0.125

$$\Sigma f = 1.0 \qquad\qquad \Sigma f = 1.0$$
$$\bar{W} = \Sigma(fw)/\Sigma f = 0.95 \qquad \bar{W} = 0.75$$
$$V_w = \Sigma f(w - \bar{W})^2 /\Sigma f \qquad V_w = 0.03125$$
$$= \Sigma fw^2 - \bar{W}^2 \text{ or } 0.01125$$

The p and q values are now 0.924 and 0.076 in population 1 and 0.5833 and 0.4167 in population 2. If selection continues in the same pattern for a second generation, the $\bar{W}$ of populations 1 and 2 is found to be 0.96124 and 0.79170, respectively. The difference in genotype frequencies determines $\Delta\bar{W}$. The change of $\bar{W}$ is less at the low q value, namely at the lower V_w value. A relationship exists, therefore, between $\Delta\bar{W}$ and V_w and Fisher identifies the association as the Fundamental Theorem of Natural Selection: "The rate of increase in fitness of any organism at any time is equal to its genetic variance in fitness at that time." The above "populations" can be used to illustrate this relationship after developing the following expression for $\Delta\bar{W}$. If three genotypes can be identified as g_1, g_2 and g_3, the following notation exists:

First Period

Genotypes	g_1	g_2	g_3
Frequencies Before-Selection	f_1	f_2	f_3
Fitnesses	w_1	w_2	w_3
Frequencies After-Selection	$f_1 w_1$	$f_2 w_2$	$f_3 w_3$

$$\Sigma f = 1.0 \text{ and } \bar{W}_1 = \Sigma(fw)/\Sigma f.$$

Second Period

Genotypes	g_1	g_2	g_3
Frequencies Before-Selection	$f_1 w_1/\bar{W}_1$	$f_2 w_2/\bar{W}_1$	$f_3 w_3/\bar{W}_1$
Fitnesses	w_1	w_2	w_3
Frequencies After-Selection	$f_1 w_1^2/\bar{W}_1$	$f_2 w_2^2/\bar{W}_1$	$f_3 w_3^2/\bar{W}_1$

$$\Sigma(fw)/\overline{W}_1 = \frac{1}{\overline{W}_1} \; \Sigma(fw) = \Sigma f = 1.0$$

$$\overline{W}_2 = \frac{\Sigma(fw^2)/\overline{W}_1}{\Sigma f} = \Sigma(fw^2)/\overline{W}_1$$

$$\Delta\overline{W} = \overline{W}_2 - \overline{W}_1 = \Sigma(fw^2)/\overline{W}_1 - \overline{W}_1 = (\Sigma fw^2 - \overline{W}_1{}^2)/\overline{W}_1,$$

$$\text{or } \Delta\overline{W} = V_w/\overline{W}_1.$$

Unless the fitness values are adjusted to give $\overline{W}_1$ a value of one, $\Delta\overline{W}$ is seen to have a proportional relation to V_w rather than being directly equal to it. Current literature uses this form (Crow and Kimura, 1970; O'Donald, 1970a, b). In the above example $\Delta\overline{W}$ is 0.0112 in population 1 and 0.042 in population 2, thus numerically illustrating greater $\Delta\overline{W}$ values at higher V_w values. This example uses a single-gene locus with two alleles expressing incomplete dominance. Note that the expression $\overline{W}_2 - \overline{W}_1$ is positive, indicating an increase in mean fitness. This maximization of fitness occurs whether selection is either directional or stabilizing for a given trait; however, dominance, genotype frequency, and simultaneous consideration of several loci involve complications for which the theorem is not strictly valid (Ewens, 1969; Arunachalam and Owen, 1971).

O'Donald (1970a, b) has expressed response to selection by a modification of this theorem; namely, the change in mean relative fitness or proportionate increase in fitness, $\Delta\overline{W}/\overline{W}$. This value is equivalent to $V_w/\overline{W}^2$. He suggests that $\Delta\overline{W}/\overline{W}$ best describes response when the optimal class is in question. If a polygenic system is only partially expressed phenotypically, yet fitness distinguishes the full system, then the true optimal class may not be observed. If the frequency distribution of a trait approximates a normal or symmetrically shaped curve, then the optimal phenotype will probably occur within this observed range. An unlikely set of assumptions would be required as an alternative. On the other hand, if the lower observed unit on the phenotypic scale is zero and the distribution is clearly skewed toward the lower values, the possibility exists that an unexpressed part of the polygenic system exists that is nonetheless recognized by fitness. A possible example concerns the distribution of spot number occurring on the hind wing of the butterfly *Maniola jurtina*. A large mode is frequently seen at zero spots, with decreasing frequencies as spot number increases to five. The possibility therefore exists that negative (but not observable) spot number completes the distribution. The term $X_0 - \overline{X}$, derived by O'Donald, estimates X_0 if the polygenic system includes it, namely for stabilizing selection in which less-fit phenotypes occur above and below the optimum. As opposed to a system where some stabilizing selection operates, linear complete or pure directional selection has $\bar{X}$ less than the optimum, and no individuals existing in greater scale value than the optimum. If one assumes further that the true optimum may occur at a greater value, i.e., beyond the polygenic system's expressed range, extrapolation of the optimum's value may be impossible. The term I under this unlikely condition has no meaning, but $\Delta\bar{W}/\bar{W}$ may measure the relative increase in fitness. Recombination within a polygenic system may generate new classes and thus expand the upper limit toward a hypothetical optimum, but for any one generation the maximal fitness within existing classes is the important value. Granting, however, an incomplete expression of the existing system, the X_O may not be observed. For instance, if the trait is spot number, values below zero are not seen but the term $X_O - \bar{X}$ may estimate it with a negative sign.

Recognizing the above complications that may occur for the optimal class, O'Donald (1970a, b) develops the following expressions for $\Delta\bar{W}/\bar{W}$. The possible differences in distribution shape for the sample data require a complex expression for $\Delta\bar{W}/\bar{W}$ similar to his rationale in obtaining I^O. For the case of linear, directional selection, and letting x be the variable before-selection,

$$\Delta\bar{W}/\bar{W} = (\Delta\bar{X})^2/V_x.$$

For stabilizing selection with quadratic fitness operating about the optimum, though the optimum need not be $\bar{X}$, the general expression is:

$$\Delta\bar{W}/\bar{W} = \frac{U_4 - V_x^2 + 4V_x(X_O - \bar{X}^2) - 4U_3(X_O - \bar{X})}{[\phi - (X_O - \bar{X})^2 - V_x]^2}$$

using notations identified above for I^O. If x is normally distributed before selection, the term simplifies to:

$$\Delta\bar{W}/\bar{W} = \frac{2V_x^2 + 4V_x(X_O - \bar{X})^2}{[\phi - (X_O - \bar{X})^2 - V_x]^2}$$

and if $X_O = \bar{X}$ in addition, the expression becomes $1/2 \, (1 - V_{x_A}/V_{x_B})^2$. Like I, $\Delta\bar{W}/\bar{W}$ may also be calculated directly from sample $\bar{W}$ and V_w values, i.e., $\Sigma fw/\Sigma f$ and $\Sigma f(w - \bar{W})^2/\Sigma f$. Where $\Delta \bar{W}/\bar{W}$ estimates are available for corresponding I values, they appear in Table 3.

The $\bar{W}$ and V_w for the hypothetical data in Figure 2 above are 0.8131 and 0.0253, respectively, giving $V_w/\bar{W}^2$ as 0.0383. The histograms do not appear unduly skewed and $\Delta\bar{X}$ is only −0.101, yet $1/2(1 - V_{x_A}/V_{x_B})^2$ equals 0.0178, or less than half the direct estimate. Such differences are less with the general expression for $\Delta\bar{W}/\bar{W}$. Recall Dowdeswell's (1961) demonstration of selection on spot number in the butterfly *Maniola jurtina*. The spots vary from 0 to, rarely, 5, and involved only a range of 0 to 4 in his study. If the last three classes are grouped, the data are as follows, involving a distinctly skewed distribution:

Spot Number x	Numbers Before-Selection	Numbers After-Selection	Relative Fitness
0	124	129	1.0
1	67	111	0.69875
2,3,4	46	66	0.605146^a

[a]Fitness obtained from mean relative values of last 3 classes.

The $\bar{W}$ and V_w are 0.8382 and 0.02985, respectively, and $V_w/\bar{W}^2$ is 0.0425. O'Donald (1970a, b) gives comparable values of 0.0423 and 0.0412 by the stabilizing and directional selection expressions above. Cases like *Maniola jurtina* probably involve stabilizing selection if the distribution shape shows no meaningful change with successive generations.

If we recall that Δq varies largely depending on q, even though s is constant, then we can see how $\Delta \bar{W}$ will depend as much or more on the q value as on the s value. As shown in Fisher's theorem, selection tends to maximize fitness, but as q decreases under selection, the variance of fitness, V_w, decreases, and $\Delta \bar{W}$ becomes progressively smaller. Also recall that $\bar{s}$ or I is a mean for all classes, and as the suboptimal classes decrease in frequency, the $\bar{s}$ would theoretically decrease. So both $\Delta \bar{W}/\bar{W}$ and $\bar{s}$ or I become smaller with continued selection since they are in turn determined by the frequency of genotypes. If fitness differences exist for only the environmental component of variation, no change need occur in $\Delta \bar{W}/\bar{W}$ over successive generations because the basis of phenotypic frequencies is not modified. Since the variation of any trait will probably have both genetic and environmental components in unknown proportions, a two-generation estimate of $\Delta \bar{W}/\bar{W}$ is difficult to interpret. If the environment has changed during consecutive generations, the meaning of consecutive $\Delta \bar{W}/\bar{W}$ values remains unclear. If a significant part of the total variation is due to the environment, then $\bar{s}$ may also not decrease over consecutive generations. A single estimate of $\bar{s}$ or I will reveal the loss due to selection during the given generation but the impact on the population's genetic structure is not revealed.

If selection has been operating consistently on a trait for many generations, then we may assume that many of the determining loci are close to homozygosity. Namely, little response to continued selection would be possible without some revision of the loci concerned. A comparison of selectability, namely response to selection, for two attributes could reveal the orders of magnitude to which each type had been exposed in selection. If all genes were homozygous, then any variation in the trait would be due to environmental causes. Thus, to measure the potential for selection response one requires an estimate of that part of the observed variation arising from genetic variation. Variation due to genetic differences is due to different alleles at the causative loci. As mentioned earlier, in polygenic systems the variation that results from dominance is apparently less than the variation arising from additive gene action. The proportion of total phenotypic variance, V_P, that can be explained by additive gene action, V_A, is the most useful measure. In the terminology of quantitative genetics, this unit is the heritability or V_A/V_P. The term does not mean that a trait is inherited, but identifies that part of its variability (specifically its variance) that includes additive gene action. The higher the heritability, the higher is the likelihood of a response to selection. The portion of total variance resulting from all genetic causes, V_G/V_P, is often called heritability in the broad sense, but lacks the predictive power that V_A/V_P possesses in relation to prospects of selection response. We may conclude that if a trait has experienced strong selection, theoretically its heritability should be low.

The procedure for estimating heritability has been well analyzed by quantitative geneticists and reviewed in detail by Falconer (1960). Basically, the methodology consists of comparing the similarity of a trait's expression between relatives. The symbol for heritability in common use is h^2, and comes from the symbol h used in the early 1920's by Sewall Wright for the ratio of standard deviations. The square sign results from heritability involving variances. The most accurate estimates appear to have been taken from parent-offspring comparisons. Falconer shows heritability to equal 2 times the regression coefficient of one parent and offspring, $2 b_{op}$, or to equal the regression coefficient directly if the mean of both parents is used. For the one-parent analysis, the male is the preferred parent in order to avoid possible maternal effects.

An interesting correlation exists between characters with lower heritabilities and reproductive success, because they are often most closely associated. For instance, values for tail length and bristle number in house mice and *Drosophila melanogaster* are 0.6 and 0.5, and 0.95 for color spotting in cattle. Values of 0.15, 0.2, and 0.01 occur for litter size, egg production, and conception rate in these species, respectively (Falconer, 1960). Robertson (1955) recognizes "fitness" characters and "peripheral" characters as traits with low and high heritabilities, respectively.

The relation between heritability and fitness is not completely understood; however, heritability may be modified by selection, and the selection limit develops when all genes contributing to the favored phenotype are fixed (Falconer, 1960). Thus, when the selection limit is reached for a given gene pool, the observed variance of a trait is predominantly environmental and a low estimate of heritability results. Perhaps fitness characters have responded to selection to the degree that all favorable genes are largely fixed and only mutation or redistribution of functions between existing genes can generate further selectability. Heritability estimates reveal another interesting point. Recall that in *Maniola jurtina* the male possessed a much lower heritability than the female. Perhaps dominance exists in the genetic background of one sex more so than in the other. Also, a maternal effect may be involved. Such an event would slow the rate of increase for homozygosity in polygenic systems.

SELECTION EFFECT

If selection acts against a recessive genotype, q^2, namely, $p^2 + 2pq + q^2(1 - s)$, then reduction in genetic fitness is sq^2, often termed the substitutional component of the genetic load, a subject thoroughly discussed by Wallace (1970). Since the total of all genotype frequencies after selection is $1 - sq^2 = \bar{W}$, then $1 - \bar{W} = sq^2 = \bar{s}$, or the percentage lost by selection. If the frequency of a suboptimal type for a quantitative trait is P, then $(1 - \bar{W}) = sP$, or the percentage lost, where s is the specific selection coefficient acting directly against the suboptimal type. The specific s value is the percentage lost from the suboptimal class, while the $\bar{s}$ value is the percentage lost from the total. If W_o is given as 1.0, then the specific s for suboptimals times the suboptimal frequency gives the percentage lost by selection from the total. For instance, in the example in Figure 2, the specific s for suboptimals is 0.2305 (computed below) and P equals 0.8105; thus $(0.2305)(0.8105) = 0.1869$, or the $\bar{s}$ (percentage lost) computed above. When $\bar{s}$ is obtained in this way, it is sometimes identified as the selection effect or pressure, E, but is seen to be $\bar{s}$. Since $E, \bar{s}$, and I are small when suboptimal frequency is small, the specific s alone does not reflect a population's loss by selection. Strickberger (1968) computes E without converting survival to relative figures. If S_o and S are survival percentages of the optimal and suboptimal classes, respectively, he defines E as $(S_o - S)P$. For unadjusted values in the Figure 2 example, the selection effect would be $(120/180 - 395/770)0.8105 = 12.4$ percent; and for adjusted values, the effect is $(1.0 - [295/770/120/180])0.8105 = 18.69$ percent. The direct difference between before- and after-selection frequencies is thus not equivalent to $\bar{s}$ as it is commonly used.

All of a generation's loss cannot be explained by selection. Referring again to the Figure 2 example, recall that a value of 0.8131 was found for

$\overline{W}$, so (950)(0.8131) gives 772.5 expected numbers following selection, but we observed only 515 or 0.5421 of the original 950. Thus, the $\overline{W}$ of 0.8131 is only a part of total fitness. The remaining loss is proportionately equal for all classes and equals 0.666̄, or the loss observed for the optimal group. Overall fitness has, therefore, a phenotypic fitness $\overline{W}$ (where fitness recognizes variability), and a second factor, b, termed here ecological fitness. Since we envision that all loss by phenotypic fitness affects only the suboptimal phenotypes, we can compute a direct estimate of s for that group. The suboptimal before-selection total of 700 lost 177.5 because of selection (loss after adjusting distribution to give the optimal class a phenotypic fitness value of 1.0). The percent lost of 0.1868 is $\bar{s}$ (Σf) for the total sample or, expressed for only suboptimals, is s (frequency of suboptimals). In the first case, $s = 0.1868/\Sigma f$ or 0.1868; in the second case, $s = 0.1868/0.8105$, or 0.2305. The latter s is of course the mean value among suboptimals only and is naturally higher than the value that includes the optimal class. The specific fitness of suboptimals is thus 0.7695 or $(1.0 - 0.2305)$ and the relations are tabulated below:

	Before-Selection			After-Selection		Final
	f	N	w	wN	b	wbN
Total Sample	1.0	950	0.8131	772.5	0.666̄	515
Optimal Class	0.1895	180	1.0	180.0	0.666̄	120
Suboptimal Class	0.8105	770	0.76948	592.5	0.666̄	395

Thus, the total percentage of survival is $\overline{W}b$, or, here, $(0.8131)(0.666̄) = 0.5421$. The intensity estimate of $\bar{s}$ giving a W closest to the direct calculation is I^H of 0.111, ($\overline{W} = 0.8890$). Since 177.5 equals numbers lost by selection from a total of 435 individuals lost, then 40.8 percent of total loss was due to selection. The term selection effect seems more appropriately applied to this percentage. Note that a low selection effect may be obtained if frequency of the disadvantageous morph is low or if s is small. Thus, selection effect equals (number loss by selection)/(total number loss); however, it is unlikely that many studies can obtain the required values.

The term intensity of selection, I, has been used here to measure the proportion lost by selection and equated to a $\bar{s}$. Geneticists working with applied problems have developed a second usage. The difference between the population mean and the mean of parents chosen for breeding is the selection differential, d, and it may be either positive or negative depending on the direction of selection. Also, the selection differential will vary depending on the percentage of the population chosen for parents and the standard deviation of the population relative to the trait's variation. Thus a standardized selection differential is required, and it is expressed as d/sdp

where sdp is the standard deviation of the trait in the population. This value is known among some geneticists as the intensity of selection and has been used to estimate s values from Δq equations and the number or loci in a polygenic systems (Falconer (1960)). Such calculations require several assumptions of things unlikely to exist and have not been used in natural populations. A further usage of the term selection effect involves response to selection in the following way.

Heritability $\times$ selection differential = selection effect or response. Selection differential and heritability are as defined above. This usage has its applications in animal husbandry, agriculture, etc.

Haldane (1957) advanced a concept for expressing selection in terms of genetic deaths that consisted of judging how many individuals must be lost to bring about complete substitution at a locus by a new gene (the "Cost of Natural Selection"). A genetic death simply means an individual deprived of any contribution to the offspring of following generations. Actual deaths may not be required for this condition. If a dominant gene is to be substituted for a recessive gene, then we have:

Genotypes	AA	Aa	aa
Frequencies	p^2 + $2pq$ +		$q^2 = 1.0$
Fitnesses	1	1	$1 - s$
After-Selection Frequencies	p^2 + $2pq$ + $p^2(1-s) = 1 - sq^2$.		

The proportion lost is sq^2 and the total over the complete replacement time is Σsq^2 or D. The proportion converted to numbers is $sq^2 N$ where N is the population size. If the population size remains constant, then $\Sigma sq^2 N$ is the total required genetic deaths. Haldane obtains a solution of $\log_e p_o$ for Σsq^2 where p_o is the initial frequency of the favored gene. If p_o is 0.0002, then Σsq^2 is 8.5, and for a stable population of 10,000 each generation, a grand total of 85,000 genetic deaths is required. When incomplete dominance exists for the favored gene, DN increases, and when the favored gene is recessive, DN is greater than for the dominant substitution by approximately a factor of 10. Haldane suggests a value of $30N$ for mean loss per substitution. Other workers suggest modifications in the expression but still conclude the loss appears great (Crow and Kimura, 1970).

The analysis presents a dilemma. If such loss occurs, the number of genes undergoing simultaneous substitution is limited or the cost is excessive. The value of s simply determines the period required to accumulate the total D. Therefore, the substitution may be distributed over a large number of generations. The likelihood of the concept's validity clearly concerns population dynamics. Haldane appears to assume that b, the ecological fitness, is constant. Organisms usually have far more off-spring than required to maintain stable adult density. If phenotypic fitness, w, decreases, and ecological fitness, b, increases, then total loss could remain constant while more genetic deaths are being processed each generation. Consequently, populations may be able to pay the estimated cost without undue risk of extinction. The w and b events are shown above in sequence but of course operate simultaneously in nature. The value of b is often said to be a function of numbers; consequently, if a genetic death is not a true death, they could affect the magnitude of b.

Several processes have been identified in which population size plays an important role in selection. First, the magnitude of s necessary to overcome drift was found to be dependent on the size of the population in question. Namely, the smaller the population, the larger is the necessary value of s. Next, in a related process, if the increment to homozygous classes by the drift process is to be effective, the increase in homozygous frequency must be large enough to affect actual genotype numbers. Here, small numbers must usually be required. Thirdly, population-dependent selection was found to include both a frequency and density component. Fourth, the number lost by selection relative to total numbers lost, i.e., the selection effect, can be important in discussing the Cost of Natural Selection. An additional consideration of population numbers exists that also relates to selection effect. The low selection effect that occurs when the frequency of the suboptimal target is low has been mentioned on several occasions. Also, if the morph with highest fitness is low in fre-quency, then Δq is also small during the early generations of selection. To increase Δq during the time that the favorable type is rare by increasing s is not likely to be successful. The great majority of the population at that time is suboptimal, and the whole population could easily go to extinc-tion. Essentially, the selection effect is high and not compensated by sufficient numbers of the most-fit morph. If actual numbers of the favored type could be increased, its abundance may reach a threshold value at which selection could restructure the population for that specific type and not push the population to near extinction. In other words, a rather high selection effect could exist, but its effect on the population would be offset by the presence of sufficient numbers of a highly fit group of individuals. Fluctuations in population numbers brought on by several types of ecological phenomena set the stage for this possibility. If recom-bination gives a likelihood of 0.003 for a specific arrangement of genes resulting from the existing population genetic structure, then of 5,000 individuals only 15 will have the desirable combination, not enough to offset much selection. If a series of generations increase in numbers, then about 75 individuals of the favored type, plus the increment due to offspring resulting from previous such individuals, will be produced by the generation that reaches 25,000. Thus, a threshold can be approached by simply increasing overall population size until the favorable type exists in sufficient numbers for Δq to change rather quickly even without a large s. In this way, numerical fluctuations are thought to provide the opportunity for rapid evolution compared to the rate expected if numbers were to remain constant. Of course, fluctuations may involve only an increase in ecological fitness and no change in the genetic structure of the population will follow. Several examples are discussed by Ford (1975).

Chapter 8

Selection in Retrospect and Prospect

The nature of variation subject to natural selection was the basis for controversy in the early years following Darwin's work. These difficulties largely dissipated with the convergence of knowledge of Mendelian inheritance, of relations between genes and chromosomes, of biometry, and of inheritance of continuous variations, and recognition of mutation. A sound theoretical foundation now exists for describing expected population responses to selection for both continuous and discontinuous variation. Nonetheless, controversial issues continue to receive attention, although exchanges of opinions are less colorful than in earlier years. A major issue in recent times concerns the relative importance to evolution of Δq by nonselection forces. This question has played a large part in the search for a definition of the unit of selection. Another area in which diverse opinions developed concerns feasibility of comparing different populations for adaptiveness by $\bar{W}$ and proportions of detrimental genes by genetic loads.

The nonselection alternative for Δq with the greatest potential interest is genetic drift. The basic operation of this process was described in Chapter Two. A number of laboratory studies, mainly with *Drosophila* species, reveal the real potentiality for drift, although measured in rather artificial systems. Spiess (1962) reviews these studies and reflects the opinion of many *Drosophila* workers that drift plays an important role. Gathering data from natural populations has not produced overly convinc-

189

ing arguments, however, because several assumptions, often unstated, apply to the interpretations. These studies generally take the form of assembling a single sample or its equivalent in time from a population or units of partially subdivided populations, scoring the sample for genetic properties, and then advancing an explanation of the historical pathway leading to the observed genetic structure. An example concerns the study by Selander et al. (1971) on the deer mouse, *Peromyscus polionotus*, mentioned above. They report genetic differences, recognized by electrophoresis, for the different populations along the discontinuous Florida islands and beaches inhabited by the mice. However, ecological differences also exist between these populations. This author, having collected other kinds of animals at various seasons in the same locality, is at least convinced that the habitats are heterogeneous. Selander feels that drift best explains the genetic differences; however, the possibility seems equally strong that selection has produced the differences. Tomaszewski et al. (1973), working with ants and using the same methodology, also found a correlation of genetic structure with different habitats but chose to explain their data by selection. Both selection and drift are processes occurring through time, and genetic attributes recorded at one point on a time scale cannot be explained without a historical speculation on events such as population size, number of generations for the assumed size, degree of migration and selection, etc. The diversity of opinion on drift will probably continue. The rationales applied to a neutral unit of selection are discussed earlier but it should be clear that drift theory depends largely on such units. Van Valen (1974) provides a further outgrowth regarding neutral units of selection. The molecular structure of functionally similar proteins differs between taxonomically related groups, and the pattern of the difference has been interpreted by some workers as revealing a constant rate of evolution. The difficulty in distinguishing between a constant and an average rate has already been stressed. Van Valen suggests that natural selection, determining fitness of individuals independent of other phenomena, cannot explain the observation of constancy. He then develops a term he calls "momentary fitness." When ecological resources are plotted on a multidimensional graph giving a resource "landscape of hills and valleys," the amount of resource under the specific area occupied by a specific species is its momentary fitness. When momentary fitness increases for a species by some given value, the total momentary fitness of ecologically associated species is assumed to fall by an equal amount. Fitness changes not involving other species, such as adaptations to climate, are thus assumed to have minor importance. Van Valen has identified this concept of fitness change as the Red Queen's Hypothesis. The position for a constant molecular evolutionary rate stands in part on the observation that the number of new mutants per generation for a locus times the probability of a gene being eventually fixed equals the rate of neutral gene substitution. This rate is found to be simply the mutation rate for the gene and natural mutation rates are usually assumed to be constant. For mammalian hemoglobins, 10^9 years are judged to have transpired per codon replacement during the gene's history. If a mutation rate of 10^{-9} per codon per year exists, then the molecular substitutions constituting the differences could have been achieved by neutral processes (Crow and Kimura, 1970). The algebra of the estimate assumes similar population sizes through the period in which quite different taxa form the connecting links over very different ecological circumstances. Some molecular structures can no doubt function well over different ecological regimes but where function is in any way influenced by ecological events, the determining gene is unlikely to be neutral. The extent that molecules operate independently of the environment is unknown, but the evidence for selection is sufficiently broad to suggest that neutrality is not constant for long periods for a given gene. Likewise, as the gene moves through taxa with time, the population sizes of the different taxa will surely vary. Van Valen observes that existing concepts of natural selection can reconcile the molecular data but the reconciliation is strained. The Red Queen's Hypothesis, in the view of others, is also under strain for several reasons. This model exemplifies a line of work that seems largely destined to remain abstract. The likelihood of recognizing all important resources for a species is low, and the prospects of presenting such real data in multidimensional graphs is also low. Even if the increase of momentary fitness for a species could be measured, the model seems to require recognizing ecologically associated species and measuring their momentary fitness with the same degree of precision. These requirements appear to make the hypothesis almost untestable.

Dobzhansky (1951) in discussing fitness led several readers to believe he was using $\bar{W}$ to compare adaptiveness between different populations. Several papers were written on the issue, most of which are reprinted in Spiess (1962), and in them the force of mathematical analysis demonstrated that $\bar{W}$ is an intrapopulation measure giving no basis for comparing different populations. Wallace (1959) suggests that if different populations are to be compared, their ability to perpetuate themselves through time is the proper unit. How such a unit is best defined is not clear, and the evolutionary significance also seems unclear. Fisher (1958b) judged the issue as "a storm in a tea cup," but underlined an interesting observation emerging from the question, i.e., a general theory explaining the origin of polymophism is unlikely. The initial article was interpreted on one point as suggesting that exploitation of resources is greater in polymorphic species than monomorphic ones. The implication followed that numerous niches favored the origin of polymorphism, and it was exemplified by the fact that wide-ranging *Drosophila* species are generally more polymorphic than the more localized ones. Cain and Sheppard (1954) questioned the evidence that polymorphic species do, in fact, extract resources more efficiently and suggested that the polymorphism of wide-ranging species may explain the range rather than the reverse. Fisher (1958b) suggested polymorphism could be valuable in randomizing predation or reducing efficiency of a predator's search image. (The word 'niche' is used here to mean some unit of habitat variation.)

The possibility of comparing populations by a value denoting population fitness depends on the basic concept of fitness. In all earlier discussions, the existence of different fitnesses was recognized by the presence of different genotypes or phenotypes simultaneously in one group or population. Thus, each phenotype experienced the same mean environmental condition and their different ways of dealing with this condition led to the different fitnesses. If two populations of a species live under two different conditions, the $\bar{W}$ for each population cannot compare their relative performance. The $\bar{W}$ is simply the mean adjustment each population reaches with its own specific environment. If the two populations live together under one environmental condition, then again the comparison is meaningless, for now only one population really exists within which different types develop their own individual fitnesses. Now, if the two populations are of two different species and if they live together under the same condition, one population may well have a higher $\bar{W}$ and lower loss to selection. Unless its ecological load (in the sense discussed in Chapter Seven) offsets this advantage, it demonstrates itself to be superior in coping with the shared environment. Ecologists have long described such interactions as competitive exclusion, whereby the species with the lower $\bar{W}$, uncompensated by ecological fitness, would be evicted. In nature, of course, it is nearly impossible to recognize the boundaries between different environmental conditions or niches when all variables are considered. One may, however, observe many species living side by side and many of these species make very similar demands on the environment. Ecologists often assume that competitive exclusion operates nonetheless, and man simply cannot recognize each species' own sphere of optimal performance.

Ayala (1970) takes issue with this interpretation. Basically, he notes that the concept of competitive exclusion assumes constant fitness for the competing species and constant environmental conditions. If the $\bar{W}$ of each species population depends on its frequency relative to the competitor, the two may well coexist at an equilibrium frequency. Again, if the environment changes even slightly, and all natural environments do, then the $\overline{W}$ of each species may well reverse in relative magnitude. Thus, comparing $\overline{W}$ of populations of the same species has questionable value, whereas a realistic consideration of $\overline{W}$ for coexisting species casts doubt on the validity of competitive exclusion in nature. Perhaps one should speak instead of competitive adjustment. A somewhat similar history exists for the genetic load as for $\overline{W}$ and population comparisons. Wallace (1970) has stressed in a convincing way the fact that such loads are also intrapopulation parameters.

The multiple-locus concept as the unit of selection began to take form in the late 1960's largely because of the high levels of heterozygosity found by the method of electrophoresis. As mentioned earlier, if high levels of heterozygosity are controlled by natural selection, with each locus contributing independently to total fitness, then the total fitness may be so reduced that the fecundity of many or most species cannot pay the price. The earlier work was referred to as "single-locus population genetics" and identified as old-fashioned. The implication followed that meaningful studies of evolution could not be pursued by single-locus models. The multiple-locus direction of population genetics is well presented by Lewontin (1974). The work of earlier investigators did, however, recognize the interaction of numerous loci. For instance, Fisher's scheme for dominance modification by a constellation of modifying genes dates to the late 1920's, and linkage of the modifying loci would appear to convey a higher stability to the dominance, a notion hard to test. Again, balanced polygenic systems were clearly formulated by Mather in 1943 and led to predictions regarding linkage very similar to the ones now read of in the current literature for the multiple-locus concept. Whenever a trait's heritability is computed, the investigator recognizes genetic variation over numerous loci affecting the trait's expression. Serious heritability studies also date back to the early 1940's. The recent emphasis on multiple-locus properties associates the properties of recombination, heterozygosity, and selective loss in a more forceful way than the earlier treatments. The enthusiasm for its ability to reconcile otherwise conflicting data need not, however, cast a shadow over the study of single-locus or factor responses to selection. Many loci are observed to stabilize at frequencies distinctly different from 0.5, the value predicted if the locus is simply a component of a larger unit consisting of several linked genes. These loci may well follow independent destinies with selection. Also, single loci usually identified as factors segregating in breeding tests may actually represent the segregation of a large block of linked loci wherein we are scoring the action of only one pair. Such a segregating unit is frequently identified as a supergene. While single-locus models are indeed older, this "old-fashioned" feature hardly reduces their ability to document evolutionary processes.

The studies using electrophoresis for identifying different enzymatic proteins, and thus an inferred allelic difference for the locus coding the protein's specificity, often stress the point that loci studied in this way constitute a random sample from the full genotype. Even so, this writer knows of no locus that controls a visible polymorphism in the external phenotype, such as body color or other structural expression, whose initial gene product has also been identified by electrophoresis. The loci scored by electrophoresis code, for the greater part, enzymes of basic metabolic pathways, not processes leading to structural characters. Many of the metabolic pathways are fundamentally similar in a wide range of organisms and much genetic change could occur in a species' genotype without a need to simultaneously modify these enzymes of basic metabolism. The lack of attention to visible variation probably traces to the fact that much of the electrophoretic application to genetics started with the students of *Drosophila*. The drosophilids are a group having remarkably little visible, external variation, at least in the natural populations. This feature simply does not apply to many, perhaps a majority, of other organisms. The observation is important since visible variants are repeatedly found to be affected by selection, often involving high selection coefficients. For instance, such variants may play a major role in a species' ability to colonize adjacent habitats, thus to extend its distribution. The dark and near-white mice living on the black lava flows and the abutting white gypsum sands, respectively, in the American southwest gave good testimony of this dependency. Also, adjustments to a changing environment, such as industrial melanism of moths, depend on visible variants. The effects are not limited to predator-type selection. The latitudinal change of body size, using one of the ecological rules, reflects a likely adjustment to thermal regulation. Visible traits also play a major role in reproductive isolation between species. Courtship patterns between the sexes often entail visual clues, some structural, some behavioral. Clearly, alleles of loci inferred by electrophoresis have much to teach us about genetic systems and selection, but these loci may well omit much of a species' evolutionary change. A truly random sample of loci should include at least some loci controlling visible polymorphism but the electrophoretic process seems to exclude them.

Future studies leading to meaningful advances or requiring care in the interpretation of their importance are very largely judgements tempered by individual values. With this bit of bias, the following areas are mentioned. The number of actual estimates for s values in the field is very low. Test designs are not sensitive to small s values and we have little insight into the variability and mean of s values in relation to different units of selection. If more such data were available, perhaps requiring the adoption of other tests, the priority of questions could be better arranged. The lack of reproductive components of fitness for natural populations is clearly a handicap in evaluating the importance of differential survival. It is hoped that in the future more biologists will direct their attention toward m_x values per phenotype.

Recent years have witnessed considerable study of genetic load and the degree of heterozygosity existing in average individuals of a species. A major reason for this attention is the loss by genetic deaths that occurs when segregation affects detrimental gene combinations or genotypes at a heterozygous locus. The importance of such losses has never been clear because the ecological load or mortality, devoid of a genetic basis, has rarely been available for comparison. Such estimates seem difficult to acquire and are probably available only by studying the population in its natural setting. Until some idea of the limits to ecological loads are available, further mathematical exercises with genetic loads seem unlikely to give further insights.

The evolution of dominance is a process that remains controversial, judging from the scattered references appearing in the literature. No matter how a gene's dominance over its alternate alleles comes to exist, the experimental data reveal that the property can be modified by selection. Also, for some alleles dominance is found to vary geographically, suggesting a similar change in their accompanying modifier genes. If a recessive gene conveys some form of reduced fitness, then, theoretically, selection could reduce its frequency to low values more rapidly than it could evolve the gene's recessiveness by readjusting frequencies for alleles at an assortment of modifier loci. A simple reduction of a detrimental gene's frequency seems more economical, yet this solution is not frequently followed. Dominance is a widespread phenomenon. Pleiotrophic gene action could explain the observation since other functions may favor the recessive allele's existence. A better understanding of dominance may well be the key to explaining many types of genetic adjustments by populations. Since dominance for chromosome inversions and alleles scored by electrophoresis is less obvious or absent, *Drosophila* workers have been less concerned with the phenomenon in naturally occurring variations. Groups with more visible, natural variation, and thus more dominance, are, on the other hand, often difficult or time consuming to culture for breeding data. These technical problems have no doubt reduced the attention dominance has received. Perhaps of equal concern are forms of gene action known as variable expressivity and penetrance. Their frequency in natural variation is not clear and their relation to selection is unknown. Further insight into dominance will quite likely involve a search for pleiotrophy or variable expressions and a better understanding of the selection agents affecting the alternate phenotypes. Recognizing disadvantages is sometimes more direct than the reverse; therefore, the low-frequency phenotypes, usually the recessives, may be the clue to identifying selection agents.

The study of selection lends itself to mathematical expression of considerable sophistication. Recent years have witnessed many abstract models for various types of selection, fitness functions, etc. Such methods have played valuable roles in biology; however, in the case of selection, theory has far outpaced verification in the field and is moving, in some quarters, to a largely abstract discipline in which little contact exists with specimen or specimen-oriented biologists. It is hoped that the model makers will consider the likelihood of testing their ideas and the field biologists will gain increasing understanding of the theorist's numerical tools.

Literature Cited

Adamkewicz, S. L. 1969. Colour polymorphism in the land isopod, *Armadillidium nasatum*. Heredity 24:249–264.

Allison, A. C. 1964. Polymorphism and natural selection in human populations. Cold Spring Harbor Symp. Quant. Biol. 29:137–149.

Anderson, W. W., and T. K. Watanabe. 1974. Selection by fertility in *Drosophila pseudoobscura*. Genetics 77:559–564.

Andrewartha, H. G. 1971. Introduction to the Study of Animal Populations, 2nd Ed. Univ. of Chicago Press, Chicago. 283 p.

Antonovics, J., and A. D. Bradshaw. 1970. Evolution in closely adjacent plant populations. VIII. Clinal patterns at a mine boundary. Heredity 25:349–362.

Arnold, R. W. 1968. Climatic selection in *Cepaea nemoralis* in the Pyrenees. Philos. Trans. R. Soc. Lond. (Biol. Sci.) 253:549–593.

Arunachalam, V., and A. R. G. Owen. 1971. Polymorphisms with Linked Loci. Chapman and Hall, Ltd., London. 122 p.

Ayala, F. J. 1970. Competition, coexistence and evolution. *In* M. K. Hecht and W. C. Steere (eds.), Essays in evolution and genetics in honor of Theodosius Dobzhansky, pp. 121–158. Appleton-Century-Crofts, New York.

Bantock, C. R., and J. A. Bayley. 1973. Visual selection for shell size in *Cepaea* (Held.) J. Anim. Ecol. 42:247–261.

Bateman, A. J. 1948. Intrasexual selection in *Drosophila*. Heredity 2:349–368.

Bell, G. 1973. The reduction of morphological variation in natural populations of smooth newt larvae. J. Anim. Ecol. 43:115–128.

Benson, W. W., and T. C. Emmel. 1973. Demography of gregariously roosting populations of the nymphaline butterfly *Marpesia berania* in Costa Rica. Ecology 54:326–335.

Birch, L. C. 1948. The intrinsic rate of natural increase of an insect population. J. Anim. Ecol. 17:15–26.

Bishop, J. A. 1972. An experimental study of the cline of industrial melanism in *Biston betularia* (L.) (Lepidoptera) between urban Liverpool and rural North Wales. J. Anim. Ecol. 41:209–243.

Blair, W. F. 1947. Estimated frequencies of the buff and grey genes (*Gr, g*) in adjacent populations of deer-mice (*Peromyscus maniculatus blandus*) living on soils of different colors. Contrib. Lab. Vert. Biol., Univ. Mich. 36:1–16.

Blest, A. D. 1963. Longevity, palatability and natural selection in five species of New World saturnid moths. Nature 197:1183–1186.

Bodmer, W. F., and A. W. F. Edwards. 1960. Natural selection and the sex ratio. Ann. Hum. Genet. 24:239–244.

Boorman, S. A., and P. R. Levitt. 1972. Group selection on the boundary of a stable population. Proc. Nat. Acad. Sci. 69:2711–2713.

Brower, J. Van Z. 1958a. Experimental studies of mimicry in some North American butterflies. Part I. The Monarch, *Danaus plexippus*, and Viceroy, *Limenitus archippus archippus*. Evolution 12:32–47.

Brower, J. Van Z. 1958b. Experimental studies of mimicry in some North American butterflies. Part 2. *Battus philenor* and *Papilio froilus, P. polyxenes* and *P. glaucus*. Evolution 12:123–136.

Brower, J. Van Z. 1958c. Experimental studies of mimicry in some North American butterflies. Part 3. *Danaus gilippus berenice* and *Limenitis archippus floridensis*. Evolution 12:273–285.

Brower, J. Van Z. 1960. Experimental studies of mimicry. IV. The reactions of Starlings to different proportions of models and mimics. Amer. Nat. 94:271–282.

Brower, L. P., and J. Van Z. Brower. 1972. Parallelism, convergence, divergence and the new concept of advergence in the evolution of mimicry. Trans. Conn. Acad. Arts and Sci. 44:59–67.

Brower, L. P., J. Van Z. Brower, and P. W. Westcott. 1960. The reactions of Toads (*Bufo terrestris*) to Bumblebees (*Bombus americanorum*) and their Robberfly mimics (*Mallophora bomboides*). Amer. Nat. 94: 343–355.

Bumpus, H. C. 1896. The variations and mutations of the introduced sparrow, *Passer domesticus*. Biology Lecture, Marine Biology Lab, Woods Hole (1896–1897). pp. 1–15.

Cain, A. J., and J. D. Currey. 1963. Area effects in *Cepaea*. Philos. Trans. R. Soc. Lond. (Biol. Sci.) 246:1–81.

Cain, A. J., and J. D. Currey. 1968. Ecogenetics of a population of *Cepaea nemoralis* subject to strong area effects. Philos. Trans. R. Soc. Lond. (Biol. Sci.) 253:447–482.

Cain, A. J., and P. M. Sheppard. 1954. The theory of adaptive polymorphism. Amer. Nat. 88:321–326.

Camin, J. H., and P. R. Ehrlich. 1958. Natural selection in water snakes (*Natrix sipedon* L.) on islands in Lake Erie. Evolution 12:504–511.

Campbell, B., (ed.). 1972. Sexual Selection and the Descent of Man 1871–1971. Aldine Publ. Co., Chicago. 378 p.

Carpenter, G. D. H. 1941. The relative frequency of beak-marks on butterflies of different edibility to birds. Proc. Zool. Soc. London, Ser. A. III:223–231.

Carson, H. L. 1961. Heterosis and fitness in experimental populations of *Drosophila melanogaster*. Evolution 15:496–509.

Cavalli-Sforza, L., and W. F. Bodmer. 1971. The Genetics of Human Populations. W. H. Freeman and Co., San Francisco. 965 p.

Clarke, B. 1964. Frequency-dependent selection for the dominance of rare polymorphic genes. Evolution 18:364–369.

Clarke, B. 1966. The evolution of morph-ratio clines. Amer. Nat. 100: 389–402.

Clarke, B. 1970. Selective constraints on amino acid substitutions during the evolution of proteins. Nature 228:159–160.

Clarke, C. A., and P. M. Sheppard. 1960. The evolution of dominance under disruptive selection. Heredity 14:73–87.

Clarke, C. A., and P. M. Sheppard. 1966. A local survey of the distribution of industrial melanic forms in the moth *Biston betularia* and estimates of the selective values of these in an industrial environment. Proc. R. Soc. Lond. (Biol.) 165:424–439.

Cook, L. M. 1971. Coefficients of Natural Selection. Hutchinson and Co., London. 207 p.

Cook, L. M., and P. O'Donald. 1971. Shell size and natural selection in *Cepaea nemoralis. In* Robert Creed (ed.), Ecological Genetics and Evolution. Blackwell Scientific Publ., Oxford and Edinburgh. 391 p.

Cooper, D. W. 1968. The use of incomplete family data in the study of selection and population structure in marsupials and domestic animals. Genetics 60:147–156.

Coppinger, R. P. 1969. The effect of experience and novelty on avian feeding behavior with reference to the evolution of warning coloration on butterflies. I. Reactions of wild-caught adult blue jays to novel insects. Behaviour 35:45–60.

Coppinger, R. P. 1970. The effect of experience and novelty on avian feeding behavior with reference to the evolution of warning coloration on butterflies. II. Reactions of naive birds to novel insects. Amer. Nat. 104:323–335.

Cott, H. B. 1940. Adaptive Coloration in Animals. Methuen, London. 508 p.

Cotterman, C. W. 1954. Estimation of gene frequencies in nonexperimental populations. Statistics and Mathematics in Biology. Iowa State College Press, Ames, Iowa. 465 p.

Crosby, J. L. 1963. The evolution and nature of dominance. J. Theor. Biol. 5:35–51.

Crow, J. F. 1954. Breeding structure of populations. II. Effective population number. Statistics and Mathematics in Biology. Iowa State College Press, Ames, Iowa. 465 p.

Crow, J. F., and M. Kimura. 1970. An Introduction to Population Genetics Theory. Harper and Row, New York. 591 p.

Darlington, C. D. 1939. The Evolution of Genetic Systems. Cambridge Univ. Press, Cambridge, Massachusetts. 149 p.

Darwin, C. 1859. On the Origin of Species by Means of Natural Selection, or the Preservation of Favored Races in the Struggle for Life. John Murray, London. 479 p.

Darwin, C. 1860. The Voyage of the Beagle. John Murray, London. 524 p.

Deevey, E. S., Jr. 1947. Life tables for natural populations of animals. Q. Rev. Biol. 22:283–314.

Dice, L. R. 1947. Effectiveness of selection by owls of deer-mice (*Peromyscus maniculatus*) which contrast in color with their background. Contrib. Lab. Vert. Biol., Univ. Mich. 34:1–20.

di Cesnola, A. P. 1904. Preliminary note on the protective value of colour in *Mantis religiosa*. Biometrika 3:58–59.

di Cesnola, A. P. 1907. A first study of natural selection on *Helix arbustorum* (Helicogena). Biometrika 5:387–399.

Dobzhansky, T. 1951. Genetics and the Origin of Species. Columbia Univ. Press. New York. 364 p.

Dobzhansky, T. 1958. Genetics of natural populations. XXVII. The genetic changes in populations of *Drosophila pseudoobscura* in the American Southwest. Evolution 12:385–401.

Dobzhansky, T. 1961. On the dynamics of chromosomal polymorphism in *Drosophila*. *In* J. S. Kennedy (ed.), Symposium No. 1, Royal Ent. Society London. pp. 30–42.

Dobzhansky, T. 1970. Genetics of the Evolutionary Process. Columbia Univ. Press, New York. 505 p.

Dowdeswell, W. H. 1961. Experimental studies on natural selection in the butterfly, *Maniola jurtina*. Heredity 16:39–52.

Edmunds, M. 1974. Significance of beakmarks on butterfly wings. Oikos 25:117–118.

Edwards, J. H. 1965. The meaning of the associations between blood groups and disease. Ann. Hum. Genet. Lond. 29:77–83.

Ehrlich, P. R., and L. E. Gilbert. 1973. Population structure and dynamics of the tropical butterfly *Heliconius ethilla*. Biotropica 5:69–82.

Elton, C. 1942. Voles, Mice, and Lemmings. Problems in Population Dynamics. Clarendon Press, Oxford. 496 p.

Endler, J. A. 1973. Gene flow and population differentiation. Science 179:243–250.

Ewens, W. J. 1969. Population Genetics. Methuen & Co., Ltd., London. 147 p.

Falconer, D. S. 1960. Introduction to Quantitative Genetics. Ronald Press Co., New York. 365 p.

Fisher, R. A. 1918. The correlation between relatives on the supposition of Mendelian inheritance. Trans. R. Soc. Edinb. 52:399–433.

Fisher, R. A. 1928. The possible modification of the response of the wild type to recurrent mutations. Amer. Nat. 62:571–574.

Fisher, R. A. 1930a. The Genetical Theory of Natural Selection. Clarendon Press, Oxford. 272 p.

Fisher, R. A. 1930b. The distribution of gene ratios for rare mutations. Proc. R. Soc. Edinb. (Biol.) 50:204–219.

Fisher, R. A. 1939. Selective forces in wild populations of *Paratettix texanus*. Ann. Eugenics 9:109–122.

Fisher, R. A. 1950. Gene frequencies in a cline determined by selection and diffusion. Biometrics 6:353–361.

Fisher, R. A. 1958a. The Genetical Theory of Natural Selection. 2nd Ed. Dover, New York. 291 p.

Fisher, R. A. 1958b. Polymorphism and natural selection. J. Ecol. 46: 289–293.

Fisher, R. A., and E. B. Ford. 1947. The spread of a gene in natural conditions in a colony of the moth *Panaxia dominula* L. Heredity 1:143–174.

Fogden, M., and P. Fogden. 1974. Animals and Their Colors: Camouflage, Warning Coloration, Courtship and Territorial Display, Mimicry. Crown Publ., Inc., New York. 172 p.

Ford, E. B. 1940. Genetic research in the Lepidoptera. Ann. Eugenics 10:227–252.

Ford, E. B. 1975. Ecological Genetics. 4th Ed. Chapman and Hall, Ltd., London. 442 p.

Ford, H. D., and E. B. Ford. 1930. Fluctuations in numbers and its influence on variation in *Melitaea aurinia*. Trans. Roy. Ent. Soc. Lond. 78:345–351.

Gadgil, M. 1975. Evolution of Social Behavior through Interpopulation Selection. Proc. Nat. Acad. Sci. 72:1199–1201.

Gause, G. F. 1934. The Struggle For Existence. Williams & Wilkins, Baltimore. 163 p.

Gerould, J. H. 1921. Blue-green caterpillars. J. Exp. Zool. 34:385–412.

Goodman, O. 1974. Natural selection and a cost ceiling on reproductive effort. Amer. Nat. 108:247–268.

Grant, V. 1963. The Origin of Adaptations. Columbia Univ. Press, New York. 606 p.

Grossman, A. I., L. G. Koreneva, and L. E. Ulitskaya. 1969. Variation of the alcohol dehydrogenase locus in natural populations of *Drosophila melanogaster*. Genetika 6:91–96. (In Russian.)

Hagen, D. W., and L. G. Gilbertson. 1973. Selective predation and the intensity of selection acting upon the lateral plates of threespine stickle-backs. Heredity 30:273–287.

Haldane, J. B. S. 1924. A mathematical theory of natural and artificial selection. I. Trans. Camb. Phil. Soc. 23:19–41.

Haldane, J. B. S. 1930. A note on Fisher's theory of the origin of dominance and on a correlation between dominance and linkage. Amer. Nat. 64:87–90.

Haldane, J. B. S. 1932. The Causes of Evolution. Longmans, Green and Co., London. 235 p.

Haldane, J. B. S. 1939. The theory of the evolution of dominance. J. Genetics 37:365–374.

Haldane, J. B. S. 1942. The selective elimination of silver foxes in Eastern Canada. J. Genetics 44:296–300.

Haldane, J. B. S. 1948. The theory of a cline. J. Genetics 48:277–284.

Haldane, J. B. S. 1953. Animal populations and their regulation. New Biology 15:9–24.

Haldane, J. B. S. 1954. The measurement of natural selection. Proc. IX Int. Cong. Genet. 1:480–487.

Haldane, J. B. S. 1956. The theory of selection for melanism in Lepidop-tera. Proc. R. Soc. Lond. (Biol.) 145:303–304.

Haldane, J. B. S. 1957. The cost of natural selection. J. Genetics 55: 511–524.

Haldane, J. B. S. 1962. Evidence for heterosis in woodlice. J. Genetics 58:39–41.

Haldane, J. B. S., and S. D. Jayakar. 1963. Polymorphism due to selection of varying direction. J. Genetics 58:237–242.

Hamilton, W. D. 1964a. The genetical evolution of social behavior. I. J. Theor. Biol. 7:1–16.

Hamilton, W. D. 1964b. The genetical evolution of social behavior. II. J. Theor. Biol. 7:17–51.

Hamilton, W. D. 1966. The moulding of senescence by natural selection. J. Theor. Biol. 12:12–45.

Hamilton, W. D. 1967. Extraordinary sex ratios. Science 155:477–488.

Harding, J., R. W. Allard, and D. G. Smeltzer. 1966. Population studies in predominantly self-pollinated species. 9. Frequency dependent selection in *Phaseolus lunatus*. Proc. Natl. Acad. Sci. 56:99–104.

Harris, J. A. 1911. A neglected paper on natural selection in the English Sparrow. Amer. Nat. 45:314–318.

Hayne, D. W. 1950. Reliability of laboratory-bred stocks as samples of wild populations, as shown in a study of the variation of *Peromyscus polionotus* in parts of Florida and Alabama. Contrib. Lab. Vert. Biol., Univ. Mich. 46:1–56.

Howard, H. W. 1940. The genetics of *Armadillidium vulgare* Lafr. I. A general survey of the problems. J. Genetics 40:83–108.

Howard, H. W. 1962. The genetics of *Armadillidium vulgare* Lafr. V. Factors for body color. J. Genetics 58:29–38.

Huxley, J. S. 1939. Clines: an auxiliary method in taxonomy. Bijdr. Dierk. 27:491–520.

Johnson, C. 1964. The evolution of territoriality in the Odonata. Evolution 18:89–92.

Johnson, C. 1973. Variability, distribution, and taxonomy of *Calopteryx dimidiata* (Zygoptera: Calopterygidae). Fla. Ent. 56:207–222.

Johnson, C. 1975. Polymorphism and natural selection in ischnuran damselflies. Evol. Theory 1:81–90.

Johnson, M. S. 1971. Adaptive lactate dehydrogenase variation in the crested blenny, *Anoplorchus*. Heredity 27:205–226.

Johnston, R. F., O. M. Niles, and S. A. Rohwer. 1972. Hermon Bumpus and natural selection in the House Sparrow, *Passer domesticus*. Evolution 26:20–31.

Kettlewell, B. 1973. The Evolution of Melanism. Clarendon Press, Oxford. 423 p.

Kettlewell, H. B. D. 1956a. Further selection experiments on industrial melanism in the Lepidoptera. Heredity 10:287–301.

Kettlewell, H. B. D. 1956b. A résumé of investigations on the evolution of melanism in the Lepidoptera. Proc. R. Soc. Lond. (Biol.) 145:297–303.

Kettlewell, H. B. D. 1958. Industrial melanism in the Lepidoptera and its contributions to our knowledge of evolution. Proc. 10th Int. Cong. Ent. (1956) 2:831–841.

Kettlewell, H. B. D. 1961. Selection experiments on melanism in *Amathes glareosa* Esp. (Lepidoptera). Heredity 16:415–434.

Kettlewell, H. B. D., and R. J. Berry. 1961. The study of a cline. Heredity 16:403–414.

Kettlewell, H. B. D., and R. J. Berry. 1969. Gene flow in a cline. *Amathes glareosa* Esp. and its melanic, F. *edda* Straud. (Lep.), in Shetland. Heredity 24:1–14.

Kidwell, M. G. 1972. Genetic change of recombination value in *Drosophila melanogaster* I and II. Genetics 70:419–432 and 433–443.

Kimura, M., and G. H. Weiss. 1964. The stepping stone model of population structure and the decrease of genetic correlation with distance. Genetics 49:561–576.

Koehn, R. K., and D. J. Rasmussen. 1967. Polymorphic and monomorphic serum esterase heterogeneity in catastomid fish populations. Biochem. Genet. 1:131–144.

Krebs, C. J. 1972. Ecology. Harper and Row, New York. 694 p.

Le Moli, F. 1972. Predation on *Drosophila melanogaster* by *Scutigera coleoptrata*. Genetic origin of a disadvantageous behaviour. Acad. Nazionale dei Lincei, Ser. 8, 53:14–21.

Levene, H. 1953. Genetic equilibrium when more than one ecological niche is available. Amer. Nat. 87:331–333.

Levi, H. W. 1965. An unusual case of mimicry. Evolution 19:261–262.

Lewontin, R. C. 1974. The Genetic Basis of Evolutionary Change. Columbia Univ. Press, New York. 346 p.

Li, C. C. 1955. Population Genetics. Univ. of Chicago Press, Chicago. 366 p.

Li, C. C. 1967. Genetic equilibrium under selection. Biometrics 23:397–484.

Lloyd, J. E. 1965. Aggressive mimicry in *Photuris*: firefly femmes fatales. Science 149:653–654.

Lorenz, K. Z. 1963. On Agression. (transl. 1966) Zeitschrift für Tierpsychologie. Harcourt, Brace and World, New York. 306 p.

Malécot, G. 1959. Les modèles stochastiqnes en génétique de population. Publ. Inst. Statist. Univ. Paris 8:173–210.

Manly, B. F. J. 1972. Estimating selective values from field data. Biometrics 28:1115–1125.

Manly, B. F. J., P. Miller, and L. M. Cook. 1972. Analysis of a selective predation experiment. Amer. Nat. 106:719–736.

Mantel, N., and C. C. Li. 1974. Estimation and testing of a measure of non-random mating. Ann. Hum. Genet. 37:445–454.

Mather, K. 1943. Polygenic inheritance and natural selection. Biol. Rev. 18:32–64.

Mather, K. 1948. Nucleus and cytoplasm in differentiation. Symp. Soc. Exp. Biol. 2:196–216.

Mather, K. 1969. Selection through competition. Heredity 24:529–540.

Mather, K. 1973. Genetical Structure of Populations. Chapman and Hall, London. 197 p.

Mather, K., and J. L. Jinks. 1971. Biometrical Genetics. Cornell Univ. Press, Ithaca, New York. 382 p.

Maynard Smith, J. 1964. Kin selection and group selection. Nature 201: 1145–1147.

Maynard Smith, J. 1971. The origin and maintenance of sex. *In* G. C. Williams (ed.), Group Selection. Aldine Press. Chicago. 210 p.

Mayr, E. 1963. Animal Species and Evolution. Harvard Univ. Press, Cambridge. 797 p.

Mayr, E. 1972. Sexual Selection and Natural Selection. Sexual Selection and the Descent of Man. Aldine Publ. Co. Chicago. 378 p.

Mayr, E. 1974. The definition of the term disruptive selection. Heredity 32:404–406.

McDonald, J. F., and F. J. Ayala. 1974. Genetic response to environmental heterogeneity. Nature 250:572–573.

McWhirter, K. G. 1969. Heritability of spot-number in Scillonian strains of Meadow Brown Butterfly (*Maniola juritina*). Heredity 24:314–318.

Metcalfe, J. A., and J. R. G. Turner. 1971. Gene frequencies in the domestic cats of York: Evidence of Selection. Heredity 26:259–268.

Murray, J. 1964. Multiple mating and effective population size in *Cepaea nemoralis*. Evolution 18:283–291.

Murray, J. 1972. Genetic Diversity and Natural Selection. Hafner Publ. Co., New York. 128 p.

O'Donald, P. 1968. Measuring the intensity of natural selection. Nature 220:197–198.

O'Donald, P. 1970a. Measuring the change of population fitness by natural selection. Nature 227:307–308.

O'Donald, P. 1970b. Change of fitness by selection for a quantitative character. Theor. Pop. Biol. 1:219–232.

O'Donald, P. 1971. Natural selection for quantitative characters. Heredity 27:137–153.

O'Donald, P. 1973. A further analysis of Bumpus' data: The intensity of natural selection. Evolution 27:398–404.

O'Gower, A. K., and P. I. Nicol. 1968. A latitudinal cline of hemoglobins in a bivalve mollusc. Heredity 23:485–492.

Oldroyd, H. 1964. The Natural History of Flies. W. W. Norton, New York. 324 p.

Parsons, P. A. 1967. The Genetic Analysis of Behavior. Methuen and Co., Ltd., London. 174 p.

Parsons, P. A. 1973. Behavioural and Ecological Genetics. A Study in *Drosophila*. Clarendon Press, Oxford. 223 p.

Parsons, P. A. 1975. Male mating speed as a component of fitness in *Drosophila*. Behav. Gen. 4:395–404.

Patterson, J. T., and W. S. Stone. 1952. Evolution in the Genus *Drosophila*. Macmillan Co., New York. 610 p.

Popham, E. J. 1941. The variation in the color of certain species of *Arctocorisa* (Hemiptera, Coroxidae) and its significance. Proc. Zool. Soc. Lond. Ser. A. 111:135–172.

Poulton, E. B. 1887. The experimental proof of the protective value of colour and markings in insects in reference to their vertebrate enemies. Proc. Zool. Soc. Lond. 1887:191–274.

Powell, J. R. 1971. Genetic polymorphism in varied environments. Science 174:1035–1036.

Prout, T. 1965. The estimation of fitness from genotypic frequencies. Evolution 19:546–551.

Prout, T. 1969. The estimation of fitness from population data. Genetics 63:949–967.

Prout, T. 1971a. The relation between fitness components and population prediction in *Drosophila*. I. The estimation of fitness components. Genetics 68:127–149.

Prout, T. 1971b. The relation between fitness components and population prediction in *Drosophila* II. Population prediction. Genetics 68: 151–167.

Provine, W. B. 1971. The Origins of Theoretical Population Genetics, Univ. of Chicago Press, Chicago. 201 p.

Rasmuson, M. 1961. Genetics on the Population Level. Svenska Bokförlaget, Stockholm. 192 p.

Redfield, J. A. 1973a. The use of the incomplete family data in the analysis of genetics and selection at the Ng locus in blue grouse (*Dendragapus obscurus*). Heredity 31:35–42.

Redfield, J. A. 1973b. Demography and genetics in colonizing populations of blue grouse (*Dendragapus obscurus*). Evolution 27:576–592.

Rees, H. 1955. Genotypic control of chromosome form and behaviour. Bot. Rev. 27:288–318.

Richards, O. W. 1961. An introduction to the study of polymorphism in insects. *In* J. S. Kennedy (ed.), Symposium No. 1. Royal Ent. Society London. pp. 1–10.

Robertson, A. 1955. Selection in animals: synthesis. Cold Spring Harbor Symp. Quant. Biol. 20:225–229.

Robertson, A. 1962. Selection for heterozygotes in small populations. Genetics 47:1291 1300.

Robertson, A. 1965. The interpretation of genotypic ratios in domestic animal populations. Anim. Prod. 7:319.

Schopf, T. J., and J. L. Gooch. 1971. Gene frequencies in a marine ectoproct. A cline in natural populations related to sea temperature. Evolution 25:286–289.

Seiger, M. B. 1967. A computer simulation of the influence of imprinting on population structure. Amer. Nat. 101:47–57.

Selander, R. K., M. H. Smith, S. Y. Yang, E. W. Johnson, and J. B. Gentry. 1971. Biochemical polymorphisms and systematics in the genus *Peromyscus*. I. Variation in the old field mouse. Studies in Genetics VI. Tex. Univ. Publ. 7103:49–90.

Shapiro, A. M. 1974. Beak-mark frequency as an index of seasonal predation intensity on common butterflies. Amer. Nat. 108:229–232.

Sheppard, P. M. 1951. A quantitative study of two populations of the moth *Panaxia dominula*. L. Heredity 5:349–378.

Sheppard, P. M. 1958. Natural Selection and Heredity. Hutchinson, London. 212 p.

Simpson, G. G. 1944. Tempo and Mode in Evolution. Columbia Univ. Press, New York. 237 p.

Simpson, G. G., A. Roe, and R. C. Lewontin. 1974. Quantitative Zoology. Harcourt, Brace and Co., New York. 440 p.

Smith, D. A. 1975. Sexual selection in a wild population of the butterfly, *Danus chrysippus* L. Science 187:664–665.

Snyder, L. H. 1932. Studies in human inheritance. IX. The inheritance of taste deficiency in man. Ohio J. Science 32:436–440.

Spiess, E. B. 1962. Papers on Animal Population Genetics. Little, Brown, Boston. 513 p.

Strickberger, M. W. 1968. Genetics. Macmillan Co., New York. 868 p.

Sturtevant, A. H. 1938. On the effects of selection on social insects. Q. Rev. Biol. 13:74–76.

Sumner, F. B. 1926. An analysis of geographic variation in mice of the *Peromyscus polionotus* group from Florida and Alabama. J. Mammalogy 7:149–184.

Sumner, F. B. 1930. Genetic and distributional studies of three subspecies of *Peromyscus*. J. Genetics 23:275–276.

Thoday, J. M. 1972. Disruptive selection. Proc. Roy. Soc. Lond. (Biol.) 182:109–143.

Thoday, J. M. 1974. Definitions of disruptive selection and of interbreeding populations. Heredity 32:406–409.

Tomaszewski, E. K., H. E. Schaeffer, and F. M. Johnson. 1973. Isozyme genotype-environment associations in natural populations of the harvester ant, *Pogonomyrmex badius*. Genetics 75:405–421.

Trivers, R. L. 1972. Parental Investment and Sexual Selection. Sexual Selection and the Descent of Man. Aldine Publ. Co., Chicago. 378 p.

Turner, J. R. G., and M. H. Williamson. 1968. Population size, natural selection, and the genetic load. Nature 218:700.

Van Valen, L. 1965. Selection in natural populations. III. Measurement and estimation. Evolution 19:514–528.

Van Valen, L. 1971. Group selection and the evolution of dispersal. Evolution 25:591–598.

Van Valen, L. 1974. Molecular evolution as predicted by natural selection. J. Mol. Evol. 3:89–101.

Waddington, C. H. 1957. The Strategy of the Genes. Allen and Unwin, London. 262 p.

Wallace, B. 1959. Studies of the relative fitness of experimental populations of *Drosophila melanogaster*. Amer. Nat. 93:295–314.

Wallace, B. 1968. Topics in Population Genetics. W. W. Norton, New York. 481 p.

Wallace, B. 1970. Genetic Load: Its Biological and Conceptual Aspects. Prentice-Hall, Inc., Englewood Cliffs, New Jersey. 116 p.

Warner, R. R. 1975. The adaptive significance of sequential hermaphroditism in animals. Amer. Nat. 109:61–82.

Weldon, W. F. G. 1901. A first study of natural selection in *Clausilia laminata*. Biometrika 1:109–124.

Weldon, W. F. R. 1904. Note on a race of *Clausilia itala* (von Mortens). Biometrika 3:299–307.

White, M. J. D. 1973. Animal Cytology and Evolution. Cambridge Univ. Press, Cambridge, Mass. 961 p.

Williams, G. C. 1975. Sex and Evolution. Princeton Univ. Press, Princeton, New Jersey. 200 p.

Wilson, E. O., and W. H. Bossert. 1971. A Primer in Population Biology. Sinauer Asso., Inc. Publ., Stamford, Connecticut. 192 p.

Wolff, B. 1955. On estimating the relation between blood group and disease. Ann. Hum. Genet. 19:251–253.

Wright, S. 1929a. Fisher's theory of dominance. Amer. Nat. 63:274–279.

Wright, S. 1929b. The evolution of dominance; Comment on Dr. Fisher's reply. Amer. Nat. 63:556–561.

Wright, S. 1948. On the roles of directed and random changes in gene frequency in the genetics of populations. Evolution 2:279–294.

Wright, S., and T. Dobzhansky. 1946. Genetics of natural populations. XII. Experimental reproduction of some of the changes caused by natural selection in certain populations of Drosophila pseudoobscura. Genetics 31:125–265.

Wynne-Edwards, V. C. 1962. Animal Dispersion in Relation to Social Behaviour. Oliver and Boyd, Edinburgh. 653 p.

Wynne-Edwards, V. C. 1963. Intergroup selection in the evolution of social systems. Nature 220:623–626.

Wynne-Edwards, V. C. 1964. A rejoinder to Perrins. Nature 201:1148.

Index